STUDENT WORKBOOK

GARY S. HUVARD
Virginia Commonwealth University

RICHARD M. FELDER
North Carolina State University

ELEMENTARY PRINCIPLES OF CHEMICAL PROCESSES

THIRD EDITION

2005 Edition with Integrated Media and Study Tools

RICHARD M. FELDER
North Carolina State University

RONALD W. ROUSSEAU
Georgia Institute of Technology

WILEY

JOHN WILEY & SONS, INC.

Photo Credit: ©Rosenfeld Images, Ltd./Ken Ross Photography

To order books or for customer service, please call 1-800-CALL-WILEY (225-5945).

ISBN 978-0-471-69759-6

Printed in the United States of America.

 20 19

Printed and bound by Courier Kendallville.

Preface

This Student Workbook guides you through the solution of some of the problems in *Elementary Principles of Chemical Processes*, helping you develop and master the systematic problem-solving skills required of chemical engineers. For each problem in the Workbook, we present the strategy and steps for solving the problem. The solution steps are outlined, with blank spaces in numbered equations so you can perform some of the calculations for each problem. In the back of the Workbook, answers for some of the blank spaces are supplied so you can check yourself as you work. The pages of the Student Workbook are perforated and are meant to be torn out for submission to your instructor.

Of course, you could skip working through the problems and sidestep the calculations and just go directly to the answers we've supplied. Our strong advice to you is, don't. If you just look at our solution without trying to get it yourself first, the chances are that when you have to do something similar in the future (for example, on an exam), you won't be able to do it. On the other hand, if you struggle a bit and can't figure it out, when you finally do look at the answer you'll understand what you should have done and you'll have a much better chance of being able to do it the next time you try. And, once you grasp a calculation and can do it without help, you own it from then on! That's what this workbook is going to help you learn to do.

Gary S. Huvard
Richard M. Felder

Contents

Chapter 2
Introduction to Engineering Calculations

Name: _____

Date: _____

Chapter 2 introduces some fundamental tools needed to do engineering calculations—the basic dimensions, the units of the SI, CGS, and American engineering system, how to convert from one set of units to another, how to work with units of force and weight, accuracy and precision in numerical values, dimensional homogeneity, and some of the ways that process data is represented and analyzed. Your goal should be to practice the unit conversion skills and computational tools presented until you can carry out these calculations without reference to the text or to your notes.

PROBLEM 2.6

You are trying to decide which of two automobiles to buy. The first is American-made, costs $14,500, and has a rated gasoline mileage of 28 miles/gal. The second car is of European manufacture, costs $21,700, and has a rate mileage of 19 km/L. If the cost of gasoline is $1.25/gal and if the cars actually deliver their rated mileage, estimate how many miles you would have to drive for the lower fuel consumption of the second car to compensate for the higher cost of this car?

Strategy

In this problem you are given some information and asked to find one or more unknown quantities. In other words, it's an algebra problem. To solve it, you do the same thing you did in high school: define variables to stand for the quantities you are trying to find, convert the information in the problem statement into equations involving the unknown variables, and solve the equations. The same thing is true of most problems in the text. The problem statements will get longer and you will learn how to formulate the equations when the problems refer to chemical processes, but underneath it all, it's just algebra.

Let X equal the miles driven when the total costs for each car (purchase price plus cost of fuel) are equal. The strategy is to write an expression for each cost in terms of X, then equate the expressions and solve for X. Notice how you can build the expressions just by inserting the appropriate conversion factors (cancel the unit you don't want and replace it with the one you want), and then fill in the numbers to complete the calculations.

Solution

Let X = miles driven and C = total cost.

$$C_{American} = \$14,500 + \frac{X \text{ mi}}{} \left| \frac{\text{gal}}{28 \text{ mi}} \right| \frac{\$1.25}{\text{gal}}$$

initial cost   cost of gas to drive X miles

$$C_{European} = \$ \underline{\quad} + \frac{X \text{ (mi)}}{} \left| \underline{\hspace{5cm}} \right| \frac{\$1.25}{\text{gal}}$$ **(2.6-1)**

Equate the costs and solve for X.

$$\Rightarrow X = \underline{\quad\quad} \text{ mi}$$ **(2.6-2)**

Name: _____

Date: _____

Notes and Calculations

PROBLEM 2.9 (Variant)

Astroturf is a durable artificial surface used to cover athletic fields. A soccer field 1.00×10^2 m long by 77.1 m wide is covered with a 1/2-inch layer of Astroturf. The density of the Astroturf is 187 oz/ft^3. Calculate the weight of the Astroturf in lb$_f$ using a single dimensional equation for the calculation.

Strategy

You can calculate the volume of the Astroturf from the given dimensions, and you know that either multiplying or dividing volume by density gives you mass and multiplying mass by the acceleration of gravity gives you weight. Instead of working out the calculations individually, do it all with a single dimensional equation, starting with what you know (volume) and letting the units of the conversion factors tell you whether to multiply or divide to end up with weight in pounds. We'll give you some hints and let you fill in the rest.

Solution

We'll start by calculating the volume of the Astroturf in ft^3 from the given dimensions, converting all length and mass dimensions to ft and lb$_m$ as necessary, and go from there.

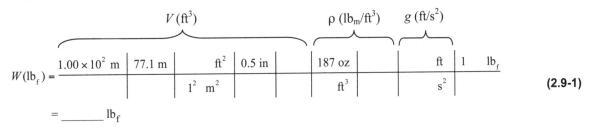

$$W(\mathrm{lb_f}) = \frac{1.00 \times 10^2 \text{ m} \mid 77.1 \text{ m} \mid \text{ft}^2 \mid 0.5 \text{ in} \mid 187 \text{ oz} \mid \text{ft} \mid 1 \mid \text{lb}_f}{1^2 \text{ m}^2 \mid \text{ft}^3 \mid \text{s}^2} \qquad (2.9\text{-}1)$$

$$= \underline{\qquad} \ \mathrm{lb_f}$$

Notes and Calculations

Name: _____

Date: _____

PROBLEM 2.25

A seed crystal of diameter D (mm) is placed in a solution of dissolved salt, and new crystals are observed to nucleate (form) at a constant rate r (crystals/min). Experiments with seed crystals of different sizes show that the rate of nucleation varies with the seed crystal diameter as

$$r \text{ (crystals/min)} = 200D - 10D^2 \qquad D \text{ in mm}$$

(a) What are the units of the constants 200 and 10? (Assume the given equation is valid and therefore dimensionally homogeneous.)

Strategy

All additive or subtractive terms in a dimensionally homogeneous equation must have the same units. Since the left-hand side has units of crystals/min, the right-hand side must have the same units, and so must each of the two terms on the right-hand side. The units of 200 and 10 are whatever they need to be to make the units of $200D$ and $10D^2$ crystals/min.

Solution

$$200D \left(\frac{\text{crystals}}{\text{min}} \right) = \frac{D(\text{mm})}{} \left| \frac{200 \text{ crystals/min}}{\text{mm}} \right. \Rightarrow 200 \frac{\text{crystals}}{\text{min} \cdot \text{mm}}$$

$$10D^2 \left(\frac{\text{crystals}}{\text{min}} \right) = \frac{D^2(\text{mm}^2)}{} \left| \frac{10 \text{ crystals/min}}{\text{mm}^2} \right. \Rightarrow 10 \frac{}{} \qquad \text{(2.25-1)}$$

(b) Calculate the crystal nucleation rate in crystals/s corresponding to a crystal diameter of 0.050 in.

Solution

Convert 0.050 in. to mm, compute r (crystals/min) by substituting the value of D (mm) into the formula, then convert the result to crystals/s. (As an exercise, try doing it with a single dimensional equation.)

(2.25-2)

```
┌─────────────────────────────────────────────────────────────────┐
│                                                                   │
│                                                                   │
│                                                                   │
│                                                                   │
│                                                                   │
│                                                                   │
└─────────────────────────────────────────────────────────────────┘
```

(c) Derive a formula for r (crystals/s) in terms of D (in.).

Strategy

Follow the procedure shown in Part 3 of Example 2.6-1 in your text. Define expressions for the original variables in terms of the new variables, then substitute the expressions in the original equation and simplify to derive the desired equation.

Solution

$$r \left(\frac{\text{crystals}}{\text{min}} \right) = \frac{r'(\text{crystals})}{s} \left| \frac{}{} \right. = 60r' \; ; \quad D(\text{mm}) = \frac{D'(\text{in})}{} \left| \frac{}{} \right. = \underline{\quad\quad} \qquad \text{(2.25-3)}$$

Substituting for r and D in $r = 200D - 10D^2$ yields

$$60r' = 200(\underline{\quad\quad}) - 10[\underline{\quad\quad\quad\quad}] \Rightarrow r' = 84.7D' - \underline{\quad\quad}(D')^2 \qquad \textbf{(2.25-4)}$$

You could now drop the primes from r' and D', recognizing that they are *not* the same as the original r and D.

Name: _____

Date: _____

PROBLEM 2.31 (Variant). Fitting non-linear (x,y) data.

State what you would plot to get a straight line if experimental (x,y) data are to be correlated by the following relations, and what the slopes and intercepts would be in terms of the parameters of the relations. If you could equally well use two different kinds of plots (e.g., rectangular or semilog), state what you would plot in each case. [The solution to part (a) is given as an example. See also Examples 2.7-2 and 2.7-3 in the text.]

(a) $y^2 = ae^{-b/x}$. (The variables are x and y, and the parameters are a and b. The goal is to find the parameter values that provide the best fit of the tabulated (x, y) data.)
   Solution: Construct a semilog plot of y^2 versus $1/x$ or a plot of $\ln(y^2)$ vs. $1/x$ on rectangular coordinates. Slope $= -b$, intercept $= \ln a$.

Strategy

This problem belongs to an important class of problems. In your engineering courses and in industrial laboratories, you may often encounter experimental data relating two variables (call them x and y), and have the task of finding a mathematical expression that fits the data. You may be given the form of an expression based on theory that includes two adjustable parameters (a and b) and be asked to calculate the parameter values that provide the best fit, or you may have to select and fit several different expressions and find the one that provides the best correlation. If x and y are not linearly related ($y = ax + b$), the strategy is to manipulate the expression algebraically into one of the form $Y = sX + I$, where X and Y are functions of the original variables (x and y) but **do not contain a and b**, and s and I are functions of the original parameters (a and b) that **do not contain x and y**. We can then calculate X and Y from the tabulated values of x and y, plot Y vs. X, determine s as the slope and and I the intercept of the linear plot, and determine a and b from s and I.

Let's illustrate this procedure with the example of part (a). You are told that two process variables, x and y, are related by the equation $y^2 = ae^{-b/x}$, and your job is to determine a and b. You set up an experiment in which you vary x, and for each value of x you measure the corresponding y. The data are as follows:

x	0.5	1.0	2.0	4.0	6.0	8.0
y	0.64	1.05	1.35	1.53	1.59	1.63

From the form of the equation, a plot of y vs. x should not be a straight line, and indeed if you plot y vs. x from the given data the graph is far from linear. (Try it.) Taking logarithms of both sides of the equation $y^2 = ae^{-b/x}$, we get

$$\underbrace{\ln(y^2)}_{Y} = \ln(a) - b/x = \underbrace{(-b)}_{s} \cdot \underbrace{\left(\tfrac{1}{x}\right)}_{X} + \underbrace{\ln(a)}_{I}$$

This equation has the desired form: X and Y are functions of x and y (which we know) but do not contain a or b (which are still unknown), and s and I are functions of a and b but not x or y. A rectangular plot of $\ln(y^2)$ vs. $(1/x)$ should therefore yield a straight line with slope $(-b)$ and intercept $\ln(a)$. Alternatively, since plotting $\ln(y^2)$ on a rectangular scale is equivalent to plotting (y^2) on a logarithmic scale, a semilog plot of y^2 (log scale) vs. $(1/x)$ (rectangular scale) would lead to the same result.

We will use the rectangular plot for the rest of the example. We augment the data table by calculating X ($= 1/x$) and Y [$= \ln(y^2)$] for each data point; plot Y vs. X; determine s and I as the slope and intercept of the resulting line, and then calculate b ($= -s$) and a [$= \exp(I)$].

x	0.5	1.0	2.0	4.0	6.0	8.0
y	0.64	1.05	1.35	1.53	1.59	1.63
$X\,(= 1/x)$	2.00	1.00	0.50	0.25	0.167	0.125
$Y\,[= \ln(y^2)]$	−0.90	0.10	0.60	0.85	0.93	0.97

⇩

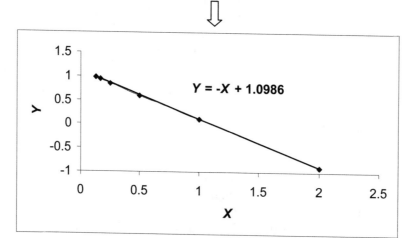

The plot turns out to be an almost perfect straight line, which means that the given nonlinear expression provides an excellent correlation of the (x,y) data. (If the plot had a distinct curvature, it would mean that either the given expression is invalid or the data are faulty.) The slope and intercept of the line yield

$$s = -1 \Rightarrow b = -s = 1 \ , \ I = 1.0986 \Rightarrow a = \exp(I) = 3.00$$
$$\Rightarrow \underline{y^2 = 3.00\,e^{-1/x}}$$

The rules for linearization are these:

1. Any combination of x and y variables may be used as an independent or dependent variable *grouping* (X or Y) as long as the resulting equation has the form of a straight line.

2. Constants may be included in an X or Y grouping *only if their values are known* (which excludes any unknown adjustable parameters). If a plot of Y vs. X is to be used to determine the value of a constant, X and Y must contain only known quantities.

For example, if

$$y^2 = 1 + \left[a(x-3)^3\right]^{-1} \Rightarrow y^2 - 1 = \left(\frac{1}{a}\right)\frac{1}{(x-3)^3}$$

one possibility would be to let $Y = (y^2 - 1)$ and $X = 1/(x-3)^3$, so that $s = (1/a)$ and $I = 0$. If the formula is valid, a plot of Y vs. X should be a straight line through the origin with a slope of $1/a$. Given a table of (x, y) pairs, you can calculate X and Y values and plot Y versus X to get a straight line with slope $s = 1/a$. You can then determine the value of the parameter a from the slope as $a = (1/s)$.

Use this procedure to determine the groupings (Y and X) that you would plot to determine the parameters a and b from experimental y vs. x data (or more generally, *dependent variable* vs. *independent variable* data) and what the slope and intercept would be in terms of the original parameters.

(b) $\sin y = a(\cos x)^2 + b$

$Y = \underline{\sin y}, \; X = \underline{\hspace{1.5cm}}, \; s = \underline{\hspace{1.5cm}}, \; I = \underline{\hspace{1.5cm}}.$ (2.31-1)

(c) $y^3 = a \ln x + \ln b$

$Y = \underline{\hspace{1cm}}, \; X = \underline{\hspace{1cm}}, \; s = \underline{\hspace{1cm}}, \; I = \underline{\hspace{1cm}}.$ (2.31-2)

(d) $C = k_1 p + k_2 \dfrac{1}{p}$ C is the dependent variable, p is the independent variable

$Y = pC, \; X = \underline{\hspace{1.5cm}}, \; s = \underline{\hspace{1.5cm}}, \; I = \underline{\hspace{1.5cm}}.$ (2.31-3)

or $Y = \underline{\hspace{1.5cm}}, \; X = \underline{\hspace{1.5cm}}, \; s = \underline{\hspace{1.5cm}}, \; I = k_1.$ (2.31-4)

(e) $\dfrac{P\hat{V}}{RT} = 1 + \dfrac{B}{\hat{V}}$ P is dependent; $\hat{V}$ is independent; T is constant

$Y = \underline{\hspace{1.5cm}}, \; X = \underline{\hspace{1.5cm}}, \; s = \underline{\hspace{1.5cm}}, \; I = 1.$ (2.31-5)

(f) $E = \dfrac{1}{2} mu^2$ E is dependent, u is independent

$Y = \underline{\hspace{1.5cm}}, \; X = \underline{\hspace{1.5cm}}, \; s = \underline{\hspace{1.5cm}}, \; I = \underline{\hspace{1.5cm}}.$ (2.31-6)

(g) $y = \dfrac{A}{B + x}$

$Y = 1/y, \; X = \underline{\hspace{1.5cm}}, \; s = \underline{\hspace{1.5cm}}, \; I = \underline{\hspace{1.5cm}}.$ (2.31-7)

(h) $C = C_0 e^{-kt}$ C is dependent, t is independent, C_0 is constant

(rect.) $Y = \underline{\hspace{1.5cm}}, \; X = \underline{\hspace{1.5cm}}, \; s = \underline{\hspace{1.5cm}}, \; I = \underline{\hspace{1.5cm}}.$ (2.31-8)

(semilog) $Y = \underline{\hspace{1.5cm}}, \; X = \underline{\hspace{1.5cm}}, \; s = \underline{\hspace{1.5cm}}, \; I = \underline{\hspace{1.5cm}}.$ (2.31-9)

(i) $p = p_0 \exp(\dfrac{-\Delta H_v}{RT})$ p is dependent, T is independent, all others constant

(rect.) $Y = \underline{\hspace{1.5cm}}, \; X = \underline{\hspace{1.5cm}}, \; s = \underline{\hspace{1.5cm}}, \; I = \underline{\hspace{1.5cm}}.$ (2.31-10)

($\underline{\hspace{1cm}}$) $Y = \underline{\hspace{1.5cm}}, \; X = \underline{\hspace{1.5cm}}, \; s = \underline{\hspace{1.5cm}}, \; I = \underline{\hspace{1.5cm}}.$ (2.31-11)

(j) $Q = At^B + 4$ Q is dependent, t is independent

(rect.) $Y = \underline{\hspace{1.5cm}}, \; X = \underline{\hspace{1.5cm}}, \; s = \underline{\hspace{1.5cm}}, \; I = \ln(A).$ (2.31-12)

($\underline{\hspace{1cm}}$) $Y = \underline{\hspace{1.5cm}}, \; X = \underline{\hspace{1.5cm}}, \; s = \underline{\hspace{1.5cm}}, \; I = \underline{\hspace{1.5cm}}.$ (2.31-13)

Notes and Calculations

Name: _____

Date: _____

PROBLEM 2.32

A **hygrometer**, which measures the amount of moisture in a gas stream, is to be calibrated using the apparatus shown here (figure on p. 36 of the text). Steam and dry air are fed at known flow rates and mixed to form a gas stream with a known water content, and the hygrometer reading is recorded; the flow rate of either the water or the air is changed to produce a stream with a different water content and the new reading is recorded, and so on. The following data are taken for y (mass fraction of water in the combined stream) vs. R (the hygrometer reading):

R	5	20	40	60	80
y	0.011	0.044	0.083	0.126	0.169

(a) Draw a calibration curve and determine an equation for $y(R)$.

(b) Suppose a sample of a stack gas is inserted in the sample chamber of the hygrometer and a reading of $R = 43$ is obtained. If the mass flow rate of the stack gas is 1200 kg/h, what is the mass flow rate of the water vapor in the stack gas?

Strategy

The information about the hygrometer, the flowchart, and the description of the experiment (and similar information in other problems) are there to acquaint you with the kinds of systems and situations you may encounter as a chemical engineer. Don't let them confuse you, however. If you carefully examine what you are being asked to do in this problem, you will see that you just need to (a) plot values of y against values of R and find an equation relating the two variables, then (b) substitute for R and calculate y.

The only way to fit an equation to data using the methods presented in this book is to find a way of plotting the data to get a straight line, then find the equation of the line. (There are more sophisticated ways involving *non-linear regression* that we won't discuss.) The simplest case occurs if the two variables are linearly related, $y = aR + b$. To determine whether they are, we will plot y vs. R and if a straight line can reasonably be drawn through the data points (which will turn out to be the case in this problem) we'll go from there.

Solution

(a) Draw a calibration curve and determine an equation for $y(R)$.

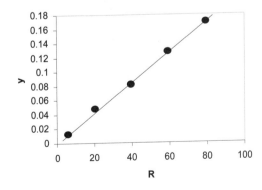

A plot of y vs. R on rectangular coordinates is a line through ($R = 5$, $y = 0.011$) and ($R = 80$, $y = 0.169$). (Two other points could equally well have been chosen, leading to slightly different results.)

$(R1 = 5, \; y1 = 0.011) \; (R2 = 80, \; y2 = 0.169)$

(2.32-1)

$$y = aR + b \Rightarrow \left. \begin{array}{l} a = \dfrac{\overline{} }{\underline{}-\underline{}} = 2.11 \times 10^{-3} \\[3mm] b = \underline{} - \left(2.11 \times 10^{-3}\right)\left(\right) = 4.50 \times 10^{-4} \end{array} \right\} \Rightarrow y = \underline{\underline{2.11 \times 10^{-3} R + 4.50 \times 10^{-4}}}$$

If the plot of y vs. R on rectangular coordinates had not been a straight line, you could either try different relationships between y and R and see if the corresponding linearized forms yield straight-line plots (as in Example 2.7-2) or plot y vs. R on semi-log or logarithmic coordinates and determine the appropriate exponential or power-law relationship if one or the other plots were linear (as in Example 2.7-3).

(b) Suppose a sample of a stack gas is inserted in the sample chamber of the hygrometer and a reading of $R = 43$ is obtained. If the mass flow rate of the stack gas is 1200 kg/h, what is the mass flow rate of the water vapor in the stack gas?

$R = 43 \Rightarrow y = $ _____ kg H_2O/kg

$\Rightarrow \dot{m}_{H_2O} = ($ _____ $)($ _____ $) = $ _____ kg H_2O/h

(2.32-2)

Name: _____

Date: _____

PROBLEM 2.38

A process instrument reading, Z (volts), is thought to be related to a process stream flow rate V (L/s) and pressure P (kPa) by the following expression:

$$Z = a\dot{V}^b P^c$$

Process data have been obtained in two sets of runs — one with $\dot{V}$ held constant, the other with P held constant. The data are as follows:

Point	1	2	3	4	5	6	7
$\dot{V}$ (L/s)	0.65	1.02	1.75	3.43	1.02	1.02	1.02
P (kPa)	11.2	11.2	11.2	11.2	9.1	7.6	5.4
Z (volts)	2.27	2.58	3.72	5.21	3.50	4.19	5.89

(a) Suppose you had only performed Runs 2, 3, and 5. Calculate a, b, and c algebraically from the data for these three runs.

Strategy

Everything is easier when you work with linear equations, so a natural strategy when you are dealing with exponential or power-law functions is to take natural logarithms. Since powers are involved in the given expression for Z, take the natural logarithm of both sides and then substitute values of $\dot{V}$, P, Z for the three data points. You will then get three equations in three unknowns (the three coefficients a, b, and c which we've bolded for clarity).

$$Z = aV^b P^c \Rightarrow \ln Z = \ln(a) + b\ln\dot{V} + c\ln P$$

using point (2): $\ln(2.58) = \ln(a) + b\ln(1.02) + c\ln(11.2)$
using point (3): $\ln(3.72) = \ln(a) + b\ln(1.75) + c\ln(11.2)$
using point (5): $\ln(3.5) = \ln(a) + b\ln(1.02) + c\ln(9.1)$

You can either solve the equations the hard way or the easy way. The hard way is to subtract (2) from (5) and find c, then subtract (2) from (3) and find b, then substitute for b and c in any one of the equations to find $\ln(a)$, then find a as $e^{\ln(a)}$. The easy way is to enter the three equations into E-Z Solve (included in the CD-ROM that came with the text) and solve them with a single mouse click. We recommend doing it the easy way partly because it's the easy way and partly because if you get practice with E-Z Solve on simple problems such as this one, you'll be ready to tackle more complex problems when you get to them (which you will). Bring up E-Z Solve, enter the following program (first filling in the missing values), choose "Solve/Sweep" under the "Solutions" menu (or just type the F5 key on your keyboard), and click on "Solve" in the dialog that appears.

(2.38-1)

E-Z Solve Code

```
// Problem 2.38(a)
ln(2.58) = lna + b*ln(1.02) + c*ln(11.2)          // "lna" is a variable name & represents ln(a)
ln(3.72) = lna + b*ln(_____) + c*ln(_____)
ln(_____) = lna + b*ln(_____) + c*ln(_____)
a = exp(lna)
```

Solution

$$a = \underline{\hspace{2cm}} \qquad b = \underline{\hspace{2cm}} \qquad c = \underline{-1.47} \qquad\qquad \textbf{(2.38-2)}$$

The value of **a** you obtain may vary a bit due to round-off error.

(b) Determine the values of **a**, **b**, and **c** using E-Z Solve and the method of least squares. Comment on why you would have more confidence in this result than in that of part (a).

Stategy

This is different from the problem in the text, which calls for a graphical solution of the sort that was common before computers made exact solutions so much easier to obtain. The graphical solution might involve the following procedure:

- Take logarithms of the expression for Z:

$$Z = a\dot{V}^b P^c \Rightarrow \ln Z = \ln\left(a\right) + b \ln \dot{V} + c \ln P$$

- Plot $\ln Z$ vs. $\ln \dot{V}$ for Runs $1 - 4$, in which P is constant. The plot would yield a straight line with slope **b** and intercept $(\ln(a) + c \ln P)$. Determine **b**. (Don't bother with the intercept.)
- Plot $\ln Z$ vs. $\ln P$ for Runs $5 - 7$, in which $\dot{V}$ is constant. The plot would yield a straight line with slope **c** and intercept $(\ln(a) + b \ln \dot{V})$. Determine **c**.
- With b and c known from the previous two calculations, Plot Z vs. $\dot{V}^b P^c$. The plot would yield a straight line through the origin with slope **a**. Determine **a**.

This procedure is cumbersome, requiring three separate plots. (The parameters could be calculated with only the first two plots by using one of the intercepts, but with less accuracy.) Instead, we can take advantage of E-Z Solve's curve-fitting capability, which works in the following way:

- Whenever you enter N algebraic equations having N unknowns into E-Z Solve, the program will return the values of the N unknowns.
- If you have N unknowns and **fewer** than N equations, E-Z Solve cannot solve the problem and will present a dialog warning you that you have too many unknowns.
- If, however, you enter **more** than N equations involving N unknowns (using the special format below), E-Z Solve will automatically enter *regression mode* and the program will estimate the best values of the N unknowns by using the Method of Least Squares. Essentially, the program (i) assumes values of **a**, **b**, and **c**, (ii) calculates the difference between each measured Z and the corresponding Z computed from the assumed **a**, **b**, and **c** values, (iii) squares each difference (or "deviation"), and (iv) sums the squares. It then repeats the procedure for different assumed values of **a**, **b**, and **c** and seeks the values that minimize the "sum of squared deviations." (Find out more about the Method of Least Squares in Appendix A of your text.)

Our strategy will therefore be to enter an equation for each run in the data set using **a**, **b**, and **c** as unknown parameters and let E-Z Solve estimate **a**, **b**, and **c** from all of the data at once.

Solution

Enter your equations into E-Z Solve as shown below, first filling in the missing values. The program will find the values of the parameters **a**, **b**, and **c** that best fit the seven data points, and will then calculate values of Z using those parameters for the same values of $\dot{V}$ and P. The calculated values will be returned as Zcalc1–Zcalc7. When all of the equations have been entered, type the F5

Name: _____

Date: _____

key and accept the default initial guesses. Click the Solve button. Record the values of *a*, *b*, and *c* returned by E-Z Solve as well as the values of Z1–Z7 calculated using those values.

E-Z Solve Code   (2.38-3)

```
// Problem 2.38(c)
//Initialize values of the dependent variables
Z1 = 2.27; Z2 = 2.58; Z3 = _____; Z4 = _____; Z5 = _____; Z6 = _____; Z7 = _____
//Force E-Z Solve to regress for values of lna, b, and c
ln(Z1) = lna + b*ln(0.65) + c*ln(11.2)
ln(Z2) = lna + b*ln(1.02) + c*ln(11.2)
ln(Z3) = lna + b*ln(1.75) + c*ln(11.2)
ln(Z4) = lna + b*ln(3.43) + c*ln(11.2)
ln(Z5) = lna + b*ln(1.02) + c*ln(9.1)
ln(Z6) = lna + b*ln(1.02) + c*ln(7.6)
ln(Z7) = lna + b*ln(1.02) + c*ln(5.4)
//
a = exp(lna)        //find a from the regressed variable lna
//
//Now calculate the auxilliary variables for comparison to the input Z values
Zcalc1 = a*0.65^b*11.2^c
Zcalc2= a*1.02^b*11.2^c
Zcalc3 = a*1.75^b*11.2^c
Zcalc4 = a*3.43^b*11.2^c
Zcalc5 = a*1.02^b*9.1^c
Zcalc6 = a*1.02^b*7.6^c
Zcalc7 = a*1.02^b*5.4^c
```

Record the values you obtain with E-Z Solve below.   (2.38-4)

a = _____	b = _____	c = -1.05				
Run	1	2	3	4	5	6	7

Run   1   2   3   4   5   6   7
Z   2.27 2.58 3.72 5.21 3.50 4.19 5.89
Zcalc 2.19

(2.38-5)

Q: Why would you have more confidence in the values of *a*, *b*, and *c* determined in part (b) than in those determined in part (a)?

A: _____

_____

Name: _____

Date: _____

Notes and Calculations

Chapter 3
Processes and Process Variables

Name: _____

Date: _____

Chapter 3 introduces a number of physical properties that characterize chemical process materials. Consistently carrying units through all calculations is a key to solving the problems in this chapter. If you write the density of liquid acetone (for example) as 791 kg/m^3 or 0.791 g/cm^3 as opposed to 0.791, you will never have to guess whether to multiply or divide by the density to convert a mass of acetone to its equivalent volume or a volumetric flow rate to its equivalent mass flow rate; just make sure the units cancel appropriately and the rest will take care of itself. Similarly, if a gas stream contains 35 mole% H_2, write the mole fraction as 0.35 mol H_2/mol rather than just 0.35 and you will always convert correctly from the total molar flow rate to the molar flow rate of hydrogen and vice versa.

PROBLEM 3.3

The specific gravity of gasoline is approximately 0.70.

(a) Determine the mass (kg) of 50.0 liters of gasoline. (Pay attention to significant figures.)

Solution

Start with the given quantity (50.0 L) and use the specific gravity (converted to density) to determine the mass.

Eqs. (3.1-1) & (3.1-2)

(3.3-1)

$$\frac{50.0 \text{ L}}{} \left| \frac{0.70 \times (1.000 \times 10^3) \text{ kg}}{\text{m}^3} \right| \frac{1 \text{ m}^3}{10^3 \text{ L}} = \underline{35 \text{ kg}}$$

(b) The mass flow rate of gasoline exiting a refinery tank is 1150 kg/min. Estimate the volumetric flow rate in liters/s.

Solution

Start with the mass flow rate and use the specific gravity to determine the volumetric flow rate. Complete the conversion equation and check your answer.

$$\frac{1150 \text{ kg}}{\text{min}} \left| \frac{\text{m}^3}{\text{kg}} \right| \frac{1 \text{ min}}{60 \text{ s}} = \underline{27.0 \text{ L/s}}$$

(3.3-2)

(c) Estimate the average mass flow rate (lb$_m$/min) delivered by a gasoline pump.

Solution

First, you need to estimate the flow rate from a gasoline pump. If you are in the United States, you should have an approximate idea of how many gallons an automobile gas tank normally holds and how long it takes to fill the tank. Use that as a starting point for your estimation, and then convert to the desired units. If you are anywhere but the U.S., start with a tank size in liters and think of a gallon as roughly four liters. Note that there is not a "correct" answer to this estimation problem: we are simply looking for a number with the right order of magnitude.

$$\frac{\underline{\quad} \text{ gal}}{\underline{\quad} \text{ min}} \left| \frac{1 \text{ ft}^3}{\underline{\quad} \text{ gal}} \right| \frac{\underline{\qquad\qquad} \text{ lb}_m}{\text{ft}^3} \cong \underline{\quad} \text{ lb}_m/\text{min}$$

(3.3-3)

(d) Gasoline and kerosene (specific gravity = 0.82) are blended to obtain a mixture with a specific gravity of 0.78. Calculate the volumetric ratio (volume of gasoline/volume of kerosene) of the two compounds in the mixture, assuming $V_{blend} = V_{gasoline} + V_{kerosene}$.

Strategy

We could begin with any arbitrary volume of one liquid (say, kerosene), calculate the volume of the other liquid (gasoline) we would need to produce a blend with the given specific gravity, and then calculate the ratio of the two volumes. The final answer would not depend on the amount initially assumed. Let us assume that we begin with a volume of 1.00 cm^3 of kerosene (we call this amount the *basis of calculation*).

Starting in Chapter 4, most problems in this book will involve taking one or more feed materials (in this problem, two of them) and producing one or more products (in this problem, one). The best way to solve most such problems is to first draw and label a picture of what's going on. The masses and volumes of the initial feeds and the final mixture may be shown as follows:

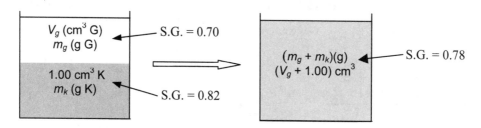

The total mass of the blend must equal the sum of the masses of the two feed liquids (why?), and the total volume, we are told, is the sum of the volumes of the feed liquids.[3.1] By definition, the ratio of the total mass to the total volume of the final mixture must equal the mixture density, which is the specific gravity (0.78) × 1.00 g/cm^3. We can use the known specific gravities to relate the masses and volumes of the feed liquids and of the final blend.

Solution

Gasoline density : $\dfrac{m_g\,(g)}{V_g\,(cm^3)} = 0.70 \times 1.00 \text{ g/cm}^3$

Kerosene density : $\dfrac{\underline{}\,(g)}{\underline{}\ cm^3} = \underline{} \times 1.00 \text{ g/cm}^3$ (3.3-4)

Blend density : $\dfrac{\underline{}\ g}{\underline{}\ cm^3} = \underline{} \times 1.00 \text{ g/cm}^3$

We now have three equations in three unknowns (m_g, m_k, and V_g). We can (a) solve the first equation for m_g, solve the second equation for m_k, substitute for the two masses in the third equation to

---

[3.1]   The assumption of "volume additivity" or "no volume change on mixing" is used often in the text and is usually a good assumption when mixing organic solvents. However, there are many mixtures, especially those containing water or other very polar substances, for which this is a very poor assumption. Examples include aqueous acids and bases. To estimate accurately a density for a mixture of such substances, you must have experimental data for the density as a function of composition. *Perry's Handbook* and the *Handbook of Chemistry and Physics* contain density data for many commonly used aqueous acids and bases.

obtain one equation in one unknown (V_g), and solve that equation for V_g or (b) let E-Z Solve do the algebra for us. The equations might be entered as

//Solution to Problem 3.3d

mg/Vg = 0.70

$$\rule{10cm}{0.4pt} \tag{3.3-5}$$

$$\rule{8cm}{0.4pt}$$

Either way, once we have V_g (cm^3 gasoline), since we know that $V_k = 1.00$ cm^3 kerosene, the requested volume ratio is simply

$$\text{Volumetric feed ratio} = \frac{V_g(\text{cm}^3 \text{ gasoline})}{1.00 \text{ cm}^3 \text{ kerosene}} = \underline{\hspace{1.5cm}} \text{ cm}^3 \text{ gasoline/cm}^3 \text{ kerosene.} \tag{3.3-6}$$

Notes and Calculations

PROBLEM 3.4 (Updated)

Assume the price of gasoline in France is approximately 1.09 euro/liter and the exchange rate is 0.818 euro/US$. How much would you pay, in dollars, for 50.0 kg of gasoline in France, assuming gasoline has a specific gravity of 0.70? What would the same quantity of gasoline cost in the United States at a rate of $1.95 per gallon?

Strategy

When you buy gasoline, you figure out the cost of your purchase by multiplying the unit cost ($/gal) by the amount (gal) you buy. In this problem, we are given the amount of the purchase in mass units (kg) so we need to convert this to volume units before proceeding with the usual cost calculation. The specific gravity (converted to density) gives us the relationship between the mass of gasoline and its volume. The other calculations needed are just unit conversions.

Solution

Since the cost in euros is given per liter of gasoline, we should convert kg of gasoline to liters of gasoline, determine the cost in euros, and then convert to dollars.

In France: $\dfrac{50.0 \text{ kg gas}}{0.70 \times \underline{\quad} \text{ kg}} \bigg| \dfrac{\text{L}}{\quad} \bigg| \dfrac{\underline{\quad\quad}}{\underline{\quad\quad}} \bigg| \dfrac{\underline{\quad\quad}}{\underline{\quad\quad}} = \$\underline{95}$ (3.4-1)

In the U.S.: $\dfrac{50.0 \text{ kg gas}}{} \bigg| \phantom{\underline{\quad\quad\quad\quad\quad\quad\quad\quad}} = \$\underline{\quad}$ (3.4-2)

Name: _____

Date: _____

Notes and Calculations

PROBLEM 3.10

Limestone (calcium carbonate) particles are stored in 50 L bags. The void fraction of the particulate matter is 0.30 (liter of void space per liter of total volume) and the specific gravity of solid calcium carbonate is 2.93.

(a) Estimate the bulk density of the bag contents (kg $CaCO_3$/liter of total volume).

Strategy

If the void volume is 0.30 L void space/L total, then the volume fraction of the solid must be 0.70 L $CaCO_3$/L total. If you carry the units and use dimensional equations, problems like this one almost solve themselves.

Solution

$$\rho_{bulk} = \frac{\rule{1cm}{0.4pt} \times \rule{1cm}{0.4pt} \text{ kg CaCO}_3}{\text{L CaCO}_3} \Bigg| \frac{\rule{1cm}{0.4pt} \text{ L CaCO}_3}{\text{L total}} = \frac{\rule{1cm}{0.4pt} \text{ kg CaCO}_3}{\text{L total}} \qquad (3.10\text{-}1)$$

(b) Estimate the weight (W) of the filled bags in newtons (N). State what you are assuming in your estimate.

Solution

$$W_{bag} = \overbrace{\rho_{bulk} V_{bag}}\, g = \frac{\rule{1cm}{0.4pt} \text{ kg CaCO}_3}{\text{L}} \Bigg| \frac{\rule{1cm}{0.4pt} \text{ L}}{} \Bigg| \frac{\rule{1cm}{0.4pt}}{} \Bigg| \frac{\rule{1cm}{0.4pt} \text{ N}}{\rule{1cm}{0.4pt} \text{ kg} \cdot \text{m/s}^2} \qquad (3.10\text{-}2)$$

$$= \rule{2cm}{0.4pt} \text{ N}$$

$$(3.10\text{-}3)$$

> **Q:** What has been assumed in this calculation?
>
> **A:** _____

(c) The contents of three bags are fed to a *ball mill*, a device something like a rotating clothes dryer containing steel or ceramic balls. The tumbling action of the balls crushes the limestone particles and turns them into a powder. (See pp. 20-31 of *Perry's Chemical Engineer's Handbook*, 7th ed.) The limestone coming out of the mill is put back into 50 L bags. Would the limestone (i) just fill three bags, (ii) fall short of filling three bags, or (iii) fill more than three bags? Briefly explain your answer.

Solution

$$(3.10\text{-}4)$$

Notes and Calculations

Name: _____

Date: _____

PROBLEM 3.17

The feed to an ammonia synthesis reactor contains 25.0 mole% nitrogen and the balance hydrogen. The flow rate of the stream is 3.00×10^3 kg/h. Calculate the rate of flow of nitrogen into the reactor in kg/h. (*Suggestion*: First, calculate the average molecular weight of the mixture.)

Strategy

Once we have the mixture molecular weight, we can determine the molar flow rate of the stream from the given mass flow rate, then the molar flow rate of nitrogen from the given mole fraction, and finally the mass flow rate of nitrogen from the molecular weight of nitrogen. The calculation of the average molecular weight is outlined following Example 3.3-3 in the text.

Solution

From Eq. (3.3-7) on p. 51 of the text,

$$\bar{M} = \frac{\underline{\quad} \text{ mol } N_2}{\text{mol}} \left| \frac{\underline{\quad} \text{ g } N_2}{\text{mol } N_2} + \frac{\underline{\quad} \text{ mol } H_2}{\text{mol}} \right| \frac{\underline{\quad} \text{ g } H_2}{\text{mol } H_2} \qquad \text{(3.17-1)}$$

$$= \underline{\quad} \text{ g/mol mixture}$$

$$\dot{m}_{N_2} = \frac{3.00 \times 10^3 \text{ kg}}{h} \left| \frac{\underline{\quad} \text{ kmol}}{\underline{\quad} \text{ kg}} \right| \frac{\underline{\quad} \text{ kmol } N_2}{\underline{\quad} \text{ kmol feed}} \left| \frac{\underline{\quad} \text{ kg } N_2}{\text{kmol } N_2} \right| = \underline{\quad} \frac{\text{kg } N_2}{h} \qquad \text{(3.17-2)}$$

Notes and Calculations

PROBLEM 3.29

A gas stream contains 18.0 mole% hexane and the remainder nitrogen. The stream flows to a condenser, where its temperature is reduced and some of the hexane is liquefied. The hexane mole fraction in the gas stream leaving the condenser is 0.0500. Liquid hexane condensate is recovered at a rate of 1.50 L/min.

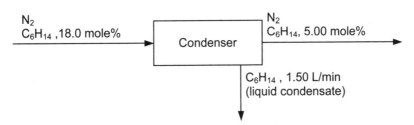

(a) What is the flow rate of the gas stream leaving the condenser in mol/min? (*Hint:* First, calculate the molar flow rate of the condensate and note that the rates at which C_6H_{14} and N_2 enter the unit must equal the total rates at which they leave in the two exit streams.)

Strategy

The hint you are given is the statement of a procedure called a *mass balance* or *material balance*. You will begin a detailed study of mass balances in Chapter 4, so this problem will be the first of many mass balance problems you encounter in this text.

As you will learn in Chapter 4, there is a systematic way to approach such problems which we will preview here. The first step is to redraw and completely label the flow diagram. The rule is that *a stream on a flowchart is completely labeled if the mass or molar flow rate of each stream component can be expressed in terms of numbers and variables labeled for that stream.* Since the mole fraction of a component times the total molar flow rate gives the molar flow rate of the component, we can achieve complete labeling by labeling the total molar flow rates and component mole fractions of the stream. A flowchart is completely labeled if every stream is completely labeled.

The flow diagram is redrawn and completely labeled below.

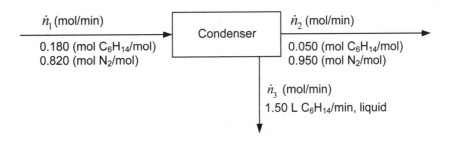

There are three unknowns on the flowchart: $\dot{n}_1, \dot{n}_2$, and $\dot{n}_3$. Since we know the volumetric flow rate of liquid hexane exiting the condenser, using the density of liquid hexane and the molecular weight of hexane (which we can look up in Table B.1 of the text) we can convert it to the molar flow rate, $\dot{n}_3$. Once $\dot{n}_3$ is known, we are left with just two unknowns. Balances on hexane and nitrogen (flow in = flow out) provide the equations needed to determine those unknowns. (The only one we need to solve Part (a) is $\dot{n}_2$.)

Solution

The specific gravity at 20°C and the molecular weight of hexane are given in Table B.1 of your text. Since we are not told the temperature of the hexane condensate, we are implicitly assuming that the specific gravity varies negligibly with temperature, a good assumption for most liquids. Use Table B.1 in the text to find the data needed and complete the calculation of $\dot{n}_3$.

Condensate flow rate (3.29-1)

$$\dot{n}_3 = \frac{1.50\ \text{L}}{\text{min}} \left| \underline{\hspace{1cm}} \times \underline{\hspace{1cm}} \frac{\text{kg C}_6\text{H}_{14}(l)}{\text{L}} \right| \frac{1000\ \text{mol}}{\underline{\hspace{1cm}}\ \text{kg}} = 11.47\ \frac{\text{mol C}_6\text{H}_{14}(l)}{\text{min}}$$

Each of the two balances that follow has the form [total flow rate in = total flow rate out], where each term has the units mol/min. When writing the balance on hexane, remember that there are two output terms—one for the gas stream and one for the liquid condensate.

Hexane Balance (3.29-2)

$$\underline{\hspace{1cm}} \frac{\text{mol C}_6\text{H}_{14}}{\text{mol feed}} \times \dot{n}_1 \left(\frac{\text{mol feed}}{\text{min}} \right)$$

$$= \underline{\hspace{1cm}} \frac{\text{mol C}_6\text{H}_{14}}{\text{mol exit gas}} \times \underline{\hspace{1cm}} \left(\frac{\text{mol exit gas}}{\text{min}} \right) + \underbrace{\underline{\hspace{1cm}} \frac{\text{mol C}_6\text{H}_{14}(l)}{\text{min}}}_{\dot{n}_3}$$

Nitrogen Balance (3.29-3)

$$0.82\dot{n}_1 = \underline{\hspace{2cm}}\ (\text{mol N}_2\ /\ \text{min})$$

The two balances must be solved simultaneously for $\dot{n}_1$ and $\dot{n}_2$. You can do it manually (the hard way) or by using E-Z Solve (the easy way). Open E-Z Solve and type the script as it appears in the block below. Use the // syntax to add documentation and units to the script. When you have the script complete, type the F5 key, accept the default initial guesses for n_1 and n_2 that appear in the F5 dialog box, and click the Solve button. After making sure your answers are physically realistic, record the answers you obtain below.

```
0.18*n1 – 0.05*n2 = 11.47        //hexane balance, mol/min
0.82*n1 – 0.95*n2 = 0            //nitrogen balance, mol/min
```

$\dot{n}_1 = $ _____ mol/min $\dot{n}_2 = $ _____ mol/min   (3.29-4)

(b) What percentage of the hexane entering the condenser is recovered as a liquid?

$$\textbf{Hexane recovery} = \frac{\dot{n}_3 [\text{mol C}_6\text{H}_{14}(l)\ \text{recovered}]}{\underline{\hspace{1cm}}(\text{mol C}_6\text{H}_{14}(l)\ \text{fed})} \times 100\% = \frac{\underline{\hspace{1cm}}}{\underline{\hspace{1cm}}} \times 100\% = \underline{\hspace{0.5cm}}\% \quad (3.29\text{-}5)$$

Name: _____

Date: _____

PROBLEM 3.42

The level of toluene (a flammable hydrocarbon) in a storage tank may fluctuate between 10 and 400 cm from the top of the tank. Since it is impossible to see inside the tank, an open-end manometer with water or mercury as the manometer fluid is to be used to determine the toluene level. One leg of the manometer is attached to the tank 500 cm from the top. A nitrogen blanket at atmospheric pressure is maintained over the tank contents.

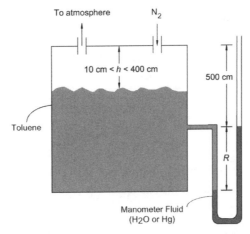

(a) When the toluene level in the tank is 150 cm below the top (h=150 cm), the manometer fluid level in the open arm is level with the point where the manometer connects to the tank. What manometer reading, R(cm), would be observed if the manometer fluid is (i) mercury, (ii) water? Which manometer fluid would you use, and why?

Strategy

Whenever you work a nontrivial manometer problem, draw a line across the manometer legs at the lowest fluid interface (see figure below and Figure 3.4-5 on p. 58 of the text) and assign variables to unknown fluid heights. Since the pressure in the manometer fluid must be the same on any horizontal plane, you can express the total pressure at the Level a in each leg in terms of the assigned variables, equate the expressions, and solve for the variable(s) of interest. You are much less likely to make a mistake this way than you would be without the sketch.

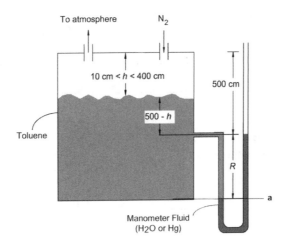

Solution

Let ρ_T be the density of toluene, ρ_M the density of the manometer fluid, and g the gravitational constant. The pressure at level **a** in the left leg of the manometer is equal to the pressure at level **a** in the right leg. Complete Eq. **(3.42-1)**. (Recall from Eq. 3.4-1 on p. 54 in the text that the pressure at the bottom of a column of fluid of density ρ and height h is the pressure at the top plus $\rho g h$, and review the discussion on p. 58.)

(3.42-1)

$$\left(P_a\right)_{left} = \left(P_a\right)_{right} \Rightarrow 1 \text{ atm} + \rho_{N_2} gh + \underline{\hspace{1cm}} + \underline{\hspace{1cm}} = 1 \text{ atm} + \rho_{air} g(500 \text{ cm}) + \underline{\hspace{1cm}}$$

The "1 atm" terms cancel and the densities of nitrogen and air are very small compared to those of the liquids (~1/1000 of the liquid densities at ambient temperature and pressure) so that we can neglect the terms involving the gas densities in the equation. Make these simplifications and solve the resulting equation for R in terms of ρ_M, ρ_T, and h.

(3.42-2)

$R =$

Substitute values for h (= 150 cm) and the densities (Table B.1) and calculate the R values for the two potential manometer fluids (mercury and water).

Densities (g/cm³): $\rho_T = \underline{\hspace{1cm}}$, $\rho_{Hg} = \underline{\hspace{1cm}}$, $\rho_{H_2O} = 1.00 \text{ g/cm}^3$ (3.42-3)

Manometer fluid is mercury $\Rightarrow R_{Hg} = \underline{\hspace{1cm}}$ cm   (3.42-4)

Manometer fluid is water $\Rightarrow R_{H_2O} = \underline{\hspace{1cm}}$ cm   (3.42-5)

So? Which manometer fluid would you use and why?   (3.42-6)

(b) Briefly describe how the system would work if the manometer were simply filled with toluene. Give several advantages of using the fluid you chose in part (a) over using toluene.

(3.42-7)

(c) What is the purpose of the nitrogen blanket? (Hint: Can you see a potential hazard associated with a mixture of air and toluene vapor?)

(3.42-8)

PROBLEM 3.47

An orifice meter is to be calibrated for the measurement of the flow rate of liquid acetone. The differential manometer fluid has a specific gravity of 1.10.

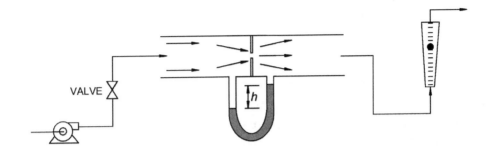

The calibration is accomplished by connecting the orifice meter in series with a rotameter (See Figure 3.2-1 and the paragraph that follows it, pg. 46-47 in the text) that has been previously calibrated for acetone (see Problem 3.13 on p. 67). A valve is adjusted to set the flow rate, whose value is determined from the rotameter reading and the rotameter calibration curve, and the corresponding differential manometer reading, h, is recorded. The procedure is repeated for several valve settings and an orifice meter calibration curve is generated by plotting flow rate versus h. The following data are taken.

Manometer Reading h (mm)	Flow Rate $\dot{v}$ (mL/s)
0	0
5	62
10	87
15	107
20	123
25	138
30	151

(a) For each of the given readings, calculate the pressure drop across the orifice, ΔP (mm Hg).

Solution

The differential manometer equation (Eq. 3.4-6 on p. 58 in your text) is $\Delta P = \left(\rho_f - \rho_{ac}\right)gh$, where ρ_f and ρ_{ac} are the densities of the manometer fluid and acetone, respectively. Complete the general equation for the pressure drop as a function of the manometer reading. Then, use a spreadsheet and complete the calibration table that follows it.

(3.47-1)

$$\Delta P(\text{mm Hg}) = \frac{(1.10 - \underline{\hspace{0.5cm}}) \text{ g}}{\text{cm}^3} \left| \frac{\underline{\hspace{0.5cm}} \text{ cm}}{\text{s}^2} \right| \frac{h \text{ (mm)}}{} \left| \frac{1 \text{ cm}}{10 \text{ mm}} \right| \frac{\underline{\hspace{0.5cm}} \text{ dyne}}{\underline{\hspace{0.5cm}} \text{ g} \cdot \text{cm/s}^2} \left| \frac{\underline{\hspace{0.5cm}} \text{ mm Hg}}{\underline{\hspace{0.5cm}} \text{ dyne/cm}^2} \right.$$

$$= \underline{\hspace{0.8cm}} \cdot h \text{ (mm)}$$

Name: _____

Date: _____

Manometer Reading h	Pressure drop ΔP (cm Hg)	Flow Rate $\dot{v}$ (mL/s)
0	0	0
5	0.114	62
10		87
15		107
20		123
25		138
30		151

Exercise

The measurement system employs a differential manometer. Without looking at the text, try to derive the differential manometer equation (Eqn. 3.4-6 on p. 58 in the text).

(b) The flow rate through an orifice should be related to the pressure drop across the orifice by the formula

$$\dot{v} = K(\Delta P)^n \qquad\qquad \text{[3.47-1]}$$

Verify graphically that the given orifice calibration data are correlated by this relationship, and determine the values of K and n that best fit the data. (If necessary, review Section 2.7d.)

Solution

If we take the logarithm of both sides of [3.47-1], we obtain the straight-line form:

$\ln \dot{v} = $ _____ + _____

(3.47-3)

The procedure for verifying that the calibration data are indeed correlated by Eq. (3.47-3) and for determining K and n is:

(3.47-4)

Plot **Y**=ln $\dot{v}$ vs. **X**=_____ on linear coordinates

Slope = _____ , Intercept = _____

Create the graph manually or (better) use a spreadsheet to

(1) compute the **X** and **Y** values and report them in your table,
(2) make the plot,
(3) fit a straight line to the data, and
(4) compute values for n and K.

Fill in the table below, print a small copy of your plot, tape it below the table, and note the values of n and K you determined in the box below. Don't forget to include the units of K.

(3.47.5)

Manometer reading h (mm)	Pressure drop ΔP (cm Hg)	Flow rate $\dot{v}$ (mL/s)	$X =$ _____	$Y = \ln \dot{v}$
0	0	0		
5	0.114	62		
10				
15				
20				
25				
30				

(3.47-6)

Orifice Calibration Curve

$n =$ _____ , $K =$ _____ (**UNITS!**)   **(3.47-7)**

(c) Suppose the orifice meter is mounted in a process line containing acetone and a reading $h = 23$ mm Hg is obtained. Determine the volumetric, mass, and molar flow rates of acetone in the line.

Solution

Find the density and molecular weight of acetone in Table B.1 in your text.

$$\Delta P = \underline{\quad} \text{ cm Hg} \qquad \dot{V} = K\left(\underline{\quad}\right)^n = \underline{\quad} \text{ mL/s} \qquad \textbf{(3.47-8)}$$

$$\dot{m} = \frac{\underline{\quad} \text{ mL}}{\text{s}} \left| \frac{\underline{\quad} \text{ g}}{\text{mL}} \right. = \underline{\quad} \frac{\text{g}}{\text{s}} \qquad \dot{n} = \frac{\underline{\quad} \text{ g}}{\text{s}} \left| \frac{1 \text{ mol}}{\underline{\quad} \text{ g}} \right. = \underline{\quad} \frac{\text{mol}}{\text{s}} \qquad \textbf{(3.47-9)}$$

Name: _____

Date: _____

Notes and Calculations

PROBLEM 3.51

A thermostat control with dial markings from 0 to 100 is used to regulate the temperature of an oil bath. A calibration plot on logarithmic coordinates of the temperature, T (°F), versus the dial setting, R, is a straight line that passes through the points ($R_1 = 20.0$, $T_1 = 110.0$°F) and ($R_2 = 40.0$, $T_2 = 250.0$°F).

(a) Derive an equation for T(°F) in terms of R.

Strategy

There is a linear relationship between the log of temperature and the log of the reading. Start by writing the form of the equation you wish to derive. Writing the intercept as ln K, as was repeatedly done in Chapter 2, is a useful trick—it simplifies conversion back to a power law after n and K have been determined.

$$\ln T\left(^\circ F\right) = n\ln\left(R\right) + \ln\left(K\right) \quad [Y = nX + \ln K]$$

A plot of **Y** [= ln(T)] vs. **X** [= ln(R)] on linear coordinates is a straight line with slope = n and intercept = ln K. The two given points on the line enable us to calculate both parameters.

Solution

Complete the calculations, noting that ($R_1 = 20.0$, $T_1 = 110.0$°F), ($R_2 = 40.0$, $T_2 = 250.0$°F)

(3.51-1)

$$n = \frac{\Delta\left(\ln T\right)}{\Delta\left(\ln R\right)} = \boxed{}$$

$$\ln(K) = \ln(T_1) - n\ln(R_1) = \ln(T_2) - n\ln(R_2) = \ln(\underline{\quad}) - \underline{\quad}\ln(\underline{\quad}) = \underline{\quad\quad} \qquad (3.51\text{-}2)$$

$$K = \underline{} \quad \text{(UNITS!)} \qquad\qquad\qquad\qquad (3.51\text{-}3)$$

Finally, write the equation in the form of a power law, $y = ax^b$ (3.51-4)

$$\boxed{}$$

(b) Estimate the thermostat setting needed to obtain a temperature of 320°F.

Solution

Solve the power law equation for R, substitute the temperature (320°F), and compute R.

$$R = \left(\frac{\boxed{}}{\boxed{}}\right)^{\boxed{}} = \underline{} \qquad (3.51\text{-}5)$$

(c) Suppose you set the thermostat to the value of R calculated in part (b) and the reading of a thermocouple mounted in the bath equilibrates at 295°F instead of 320°F. Suggest several possible explanations.

Exercise

Whenever you are faced with an unexpected result in the lab or in a process plant, the best approach to troubleshooting is to gather together a small team of people who are reasonably familiar with the experiment or process (or experiments or processes similar to it). Ask the group to think up and call out whatever ideas come to mind without criticizing or analyzing the ideas. The goal is to generate as long a list of ideas as possible—good ideas, bad ideas, even ridiculous ideas. (Sometimes, those turn out to lead to the best solutions.) When the group runs out of ideas, take a short break, return and have the group evaluate each idea and prioritize the ideas in terms of their likelihood. This process, called **brainstorming**, is often the shortest route to figuring out a puzzling problem.

Gather two or three other classmates together. Let one person take notes and all the others call out possible reasons for the observation. This process should only take 5 to 10 minutes—depending on how many funny ideas are generated. Those ideas tend to slow things down a little but they make the activity much more fun and the relaxation engendered by laughter tends to spawn creativity. After your group generates the list, take a short break, return to the problem, and try to reduce the list to those explanations the group thinks are technically the most probable. Record both lists generated by your group below.

Preliminary Ideas   (3.51-6)

Prioritized Ideas   (3.51-7)

Chapter 4
Fundamentals of Material Balances

With Chapter 4, Fundamentals of Material Balances, you begin your training as a chemical engineer. The ability of chemical engineers to use conservation laws to describe the behaviors of complex chemical and manufacturing processes is perhaps the single most important distinguishing trait of our profession. The analysis of all chemical processes begins with material balances and your mastery of this subject is critical to your success in future chemical engineering courses and your career.

Your text presents a systematic procedure for solving material balance problems (drawing and labeling a flowchart, doing a degree-of-freedom analysis, etc.). Early on, you may find some of the problems to be rather easy—in fact, so easy that you will be tempted to solve them intuitively without bothering to use that procedure. We caution you to avoid the temptation. It is vital that you learn and practice the procedure on these simple problems, since the problems will soon become much more complex and difficult and it will take you hours to solve problems that you could solve in minutes if you attacked them correctly. Later, after you have mastered the systematic approach, you may develop your own shortcuts. For now, try to turn off your intuition and work the problems *exactly* as suggested, following the specified procedural steps on every one. If you do that, you'll almost certainly pass the course and probably get a good grade as well.

Before we start going through problems and solutions, we want to offer you some additional tips on how to do well in this course. You may of course choose to skip directly to the problems. The best advice we can give you is, don't.

Success Strategies for the Stoichiometry Course

Work lots of problems in study groups
You cannot and will not learn to solve material and energy balance problems by reading a text or watching your instructor work problems on a board or an overhead projector. The *only* way to learn is by working problems yourself—lots and lots of them. We have found that students who work together in small, focused teams tend to learn how to work the problems faster than those who study alone. Working problems at a marker board or around a table, without the book and without notes, consistently leads to faster and better learning and higher grades on tests.

We therefore suggest that you form a small team with two to three other serious students. Each team member should try to set up all of the problems in an assignment *before* a team work session. The team should meet for 1–2 hours every day if possible, every other day at a minimum. During a work session, each student, in turn, goes to the marker board—no book, no notes—and completely works a problem. The others team members are allowed to give hints (sparingly) and call out any physical property data requested by the student at the board. Team members should ask questions during the solution. The idea is to make each student verbalize the solution as he or she is working the problem at the board. One other team member should record the solutions for the group, and this responsibility should rotate through the members.

When you are able to set up, solve, and verbally lead others through a problem solution, you will *own* that problem and other problems like it and will almost certainly be able to solve them on tests. This method of studying applies to all of your other courses as well. It is absolutely the best way to learn organic chemistry, for example—especially those beastly mechanisms.

Approach problems systematically
Before you go any farther in this workbook, you *must* read Chapter 4 in the text through Section 4.3 completely, and read Section 4.3d (Degree-of-Freedom Analysis) twice. (The procedure we are about to describe won't make much sense if you haven't seen it illustrated in the text, as it is in

Example 4.3-5.) As we work out problems in this workbook, we'll use the systematic, step-by-step approach given in the text. In fact, we'll add to it a bit. As new types of problems are encountered (i.e., for reactive processes), we'll modify these steps appropriately.

Note: In all the years we have taught, we cannot remember ever having a student who followed this procedure faithfully in homework problems and still failed the course. On the other hand, students who routinely skipped steps (especially Step 1) and failed the course are too numerous to count.

Here are the steps:

1. *Draw the flowchart, choose a basis of calculation, and label the flowchart completely.*

Until you have more experience deciphering problem statements and are more familiar with typical placements of process equipment, you might find it helpful to break this process down into four steps.

- *Draw the flowchart.* Read the problem statement carefully. Whenever a piece of equipment (reactor, dryer, heat exchanger, stream mixing unit, etc.) is mentioned, draw and label a box to represent it.[4.2] Then, when an input to or output from the process or a flow between units is mentioned, draw it as a line on your flowchart with an arrowhead to indicate the direction of flow, and fill in known values of amounts for batch processes (mass, moles, or volume) or mass, molar, or volumetric flow rates for continuous processes, composition variables (mass or mole fractions), and starting in Chapter 5, stream phases (gas, liquid, or solid), temperatures, and pressures.

- *Identify a basis of calculation (an amount or flow rate of a stream or one of its components).* If the problem statement specifies an amount or flow rate for one of the streams or for one of the species in a stream, it is usually convenient to choose that quantity as the basis.[4.3] If no amount or flow rate is specified, choose one yourself and write it in the appropriate location on the flowchart. Some choices lead to easier solutions, but the result will be the same no matter what you choose. The amount or flow rate of a stream with known composition is usually a good choice.

- *Complete the labeling of the flowchart by assigning variable names to unknown stream amounts or flow rates and composition variables.* The labeling is not complete until you can express the amount or flow rate of each component in every stream in terms of labeled values and variables. Label a stream with either a) the amount or flow rate of each component, or b) the total mass flow rate and component mass fractions, or c) the total molar flow rate and component mole fractions. If you label mass fractions (x_i) for N components, you are allowed

  to label N-1 of the components and let the last mass fraction (x_N) be given by $x_N = 1 - \sum_{i=1}^{N-1} x_i$,

  and similarly for mole fractions.

[4.2] If you are unsure what function a given piece of process equipment performs, *look it up and **read** about it*. You will find discussions on all the various types of equipment in *Perry's Handbook* and many different pieces of equipment are described in the *Encyclopedia of Chemical Engineering Equipment* on the CD that came with your text.

[4.3] For some problems, you may find it convenient to temporarily treat a given amount or flow rate as unknown and choose a different basis of calculation. After you have solved for all unknown variables on the flowchart, you would then scale all the stream amounts and flowrates to get back to the originally given quantity. If more than one stream amount or flow rate is specified in the problem statement, however, you must use them as the basis of calculation.

Name: _____

Date: _____

- *Make sure you can express all of the quantities requested in the problem statement in terms of the labeled quantities and variables on the flowchart, so that once you solve for the unknowns you know you will be able to complete the problem solution.*

2. *Perform a Degree-of-Freedom (DOF) Analysis.*

The purpose of this step is to make sure you have a solvable problem before you waste a lot of time trying to solve an unsolvable one. Before you begin, double-check to make sure that the flowchart is completely labeled—the procedure won't work if it isn't.

- *List and count the unknowns labeled on the flowchart.*

- *Subtract the number of independent balances allowed.* The rules for counting balances are discussed in Section 4.3c of the text. For nonreactive processes (no chemical reactions), count one balance for each independent species. (If two species always appear together in the same known proportion, they are not independent—balances on each species will reduce to the same equation.) We'll hold off on reactive processes for now.

- *Subtract 1 DOF for any known piece of information that can be used to calculate an unknown variable or any equation other than a material balance that relates unknown variables.* Such information might include:
  - process specifications (e.g., fractional conversion, a recycle ratio, a cooling rate)
  - physical properties (e.g., a component density, the relative humidity of a stream, a vapor pressure)
  - relationships between mass or molar quantities and volumetric quantities (e.g., densities of liquids and solids, equations-of-state for gases, phase equilibrium relationships)

Each time you subtract a DOF this way, list the information or relationship to be used and, in parentheses, justify the subtraction by noting the flowchart variable to be determined with the specified information or the variables related by the specified equation. You may sometimes be given extra information that cannot be used to determine unknown flowchart variables. If you count this information but have no idea which (if any) variables it involves, you run a substantial risk of wasting a huge amount of time trying to solve an unsolvable problem.

Many different types of process information will be introduced as you proceed through the text. In some cases, the information needed to solve a problem will be given in the problem statement; in others, you will have to determine when certain information is needed and, knowing that the information is available in the back of your text (or, possibly, in Perry's Handbook), look it up. The ability to recognize the different types of process information and then using each bit of information to solve for a flowchart unknown is often the most difficult part of the course to master. You will develop this skill only through a lot of practice. Try not to get frustrated early in the learning process—every chemical engineer in the world has gone through the same ordeal. You WILL learn how to do this and justifying your DOF choices will help you learn.

- *Note the number of degrees-of-freedom (n_{df}) remaining.*

If $n_{df} < 0$, you have more equations than unknowns and the problem is *overspecified*. It is possible that some of the equations are simply redundant and if you solve n of them, the solution will also satisfy the other equations; however, if you are working a problem from the text and you encounter this situation, it is safe to assume that either the flowchart is incompletely labeled or you have counted some irrelevant equations.

If $n_{df} = 0$, the problem is solvable (n equations in n unknowns).

If $n_{df} > 1$, you have more unknowns than equations and the problem is *underspecified*. Unless you have forgotten some information in the problem statement that will allow you to determine an unknown, failed to count an equation relating unknowns, *or forgotten to take a basis of calculation,* you will be wasting your time if you start writing equations hoping to solve for all of the unknowns. When you work problems on processes with multiple process units and recycle streams, try analyzing different subsystems to find one with zero degrees of freedom. *Note: None of the chapter-end problems in the text are underspecified.*

3. *Write the system equations in the order in which you will solve them.*

Write down the material balances and other equations you identified in the degree-of-freedom analysis, *without doing any algebraic or arithmetic calculations.* (You will have to fight the temptation to plunge into the algebra immediately the first few times you do this.) You should have as many equations as you have unknowns when you finish. (If you don't, look back at the DOF analysis and see what you have forgotten.)

- If you are going to solve the equations manually, write them in an order such that you have to solve as few of them simultaneously as possible, starting with equations having only one unknown if there are any. Circle the variables for which you will solve in each equation. If you have an equation with only one unknown, circle <u>that</u> unknown; if you have two equations in two unknowns, enclose the equations in curly brackets (to denote the need for simultaneous solution) and circle the two unknowns. Once you have circled an unknown, treat it as known whenever it appears in subsequent equations.

- If you plan to use E-Z Solve, the order in which you write the equations does not matter. Generally, this is true of other equation-solving software as well. However, E-Z Solve is case sensitive so don't get careless with your variable names—myx and myX are completely different variables to E-Z Solve.

4. *Solve the equations.*

If you have carried out Step 4 successfully, this part is straightforward although the algebra may be tedious. (If it looks like it will be, consider using E-Z Solve. Once you get used to it, you'll never want to do anything else.)

5. *Determine any remaining quantities requested in the problem statement.*

These might be yields, separation ratios, volumetric flow rates—just about anything extra Profs. Felder and Rousseau thought up to make the problem more realistic or interesting.

Present Legible Problem Solutions

Suppose you arrive for class one day and your professor presents the problem solution in Figure 4-1 with the remark that the class should take good notes because he plans to include a problem like it on the next test.

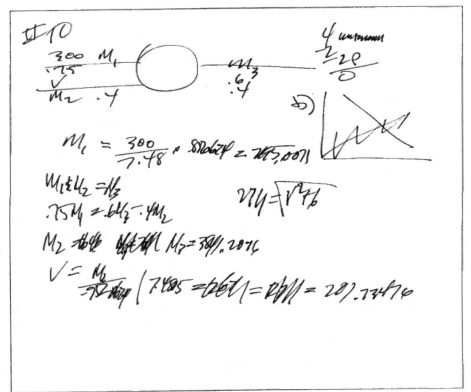

FIGURE 4.1

What's that you say? What's wrong with it? What do mean you can't figure out what he's doing? So there are no units on any of the quantities? So the graph isn't labeled. So? No, I don't know why there are so many significant figures on some of the answers. Maybe he got them with a spreadsheet.

If you find yourself snickering and thinking "What idiot would do that?" don't. Figure 4.1 is a homework assignment actually turned in by a student. When you get out of school and go to work, you will spend 35% (or more) of your time writing and sharing calculations with colleagues. Just as that professor's credibility would instantly go to zero with his class, so will yours the first time you hand a colleague a sloppy or incomprehensible set of calculations. And, that credibility rating may never improve—one event like this and you may be forever an idiot in that colleague's view. There's also another important consideration. Since the primary responsibility of the faculty is to prepare students to function as professionals, we tell our students that we will deduct major points on assignments and tests for solutions that look like that, even if the final answers are correct, and your instructor is likely to do the same.

There is a right way and many wrong ways to present chemical engineering calculations. Our hope is that you follow our lead in this workbook and learn to present your work in a legible, organized, logical fashion. Below is a short list of written presentation flaws. Try to avoid them in every assignment you hand in.

Here are the things students turn in that make us reach for our red pens while mumbling unprintable things to ourselves.

- *Messy, difficult-to-read handwriting.* Terrible penmanship is unacceptable for engineering calculations. If you cannot print plainly enough to make your writing reasonably legible, learn to typeset equations.

- *Missing or incorrect units.* Unacceptable, period.[4.1]

- *Inconsistent nomenclature* ("He wrote H—does he really mean $\hat{H}$ or $\bar{H}$? And over here, he seems to be using ϕ for volumetric flow rate but he used $\dot{v}$ before.")

- *Too many significant figures.* Turning in an answer with four significant figures when you are only entitled to two is an automatic point deduction with many instructors. Also, *nothing* irritates professors more than being handed a printout of a spreadsheet with column after column of numbers with nine digits. That level of precision is absurd, and including it in your solution instantly lowers your credibility. Formatting the numbers to a reasonable number of significant figures does not alter the precision of the calculations in any way—the spreadsheet maintains precision regardless of what is displayed.

- *Improperly or incompletely labeled flowcharts.* Do this and you've probably set yourself up to butcher the problem before you even start the calculations. Take the time to set up and correctly label the flowchart. The few minutes it takes to properly draw and label a flowchart could reduce the solution time by a factor of two or more once the problems become moderately complex.

- *Graphs without proper labeling.* Label the axes (including units!) when you make a plot. If you add a trend line to a graph on a spreadsheet, edit the default text displayed and change the "y" and "x" to the actual variables you are correlating.

- *Spreadsheets without column headings and comments.* It is difficult and time-consuming to decipher someone else's spreadsheet when the columns are not labeled and don't include units. Add comments using text blocks and briefly explain what you are doing unless it is obvious. A column labeled "ln X" next to a column labeled "X" is self-explanatory but a column labeled ψ (with no units) next to a column labeled Λ^2 requires explanation.

And that's all there is to it. Now let's do some problems.

[4.1] Simple mass balance equations (e.g., $m_1 + m_2 = m_3$) may be presented without units so long as the variables and their units have been defined on the flowchart. However, the units must be included when any variable is presented as an answer. Thus, $m_3 = 5\ \text{lb}_m$ is acceptable; $m_3 = 5$ is not.

PROBLEM 4.11

If the percentage of fuel in a fuel-air mixture falls below a certain value called the lower flammability limit (LFL), the mixture cannot be ignited. For example, the LFL of propane in air is 2.05 mole% C_3H_8. If the percentage of propane in a propane-air mixture is greater than 2.05 mole%, the gas mixture can ignite if it is exposed to a flame or spark; if the percentage is lower than the LFL, the mixture will not ignite. (There is also an upper flammability limit, which for propane in air is 11.4 mole%.)

A mixture of propane in air containing 4.03 mole% C_3H_8 (fuel gas) is the feed to a combustion furnace. If there is a problem in the furnace, a stream of pure air (dilution air) is added to the fuel mixture prior to the furnace inlet to make sure that ignition is not possible.

Note: We're going to present the solution to this problem in much greater detail than you'll see anywhere else in the workbook. Don't be intimidated by the length: we'll be using far more space to explain the steps than it will take you to carry them out when you solve similar problems. (If you don't believe it, see our hand-written worked-out solution without all the explanations at the end of the formal solution.) We strongly encourage you to read through each step carefully until you are convinced that you could do it yourself without referring to our solution. We can promise that the time it takes you to do this now will be paid back many times over when you get to your homework assignments.

(a) Draw and label a flowchart of the fuel gas-dilution air mixing unit, presuming that the gas entering the furnace contains propane at the LFL, and do the degree-of-freedom analysis.

Solution
1. Draw the flowchart, choose a basis of calculation, and complete the flowchart labeling.

At first reading, this problem might seem to be about a combustion furnace but if you examine the problem statement carefully you will find that everything described relates to things happening before the gases enter the furnace, and the furnace itself never really enters the picture. If there is a furnace malfunction, we want to flood the fuel gas with enough air to render the gas nonflammable—this happens when the propane concentration falls below the lower flammability limit, 2.05 mole%. Here is a graphical representation of what is going on.

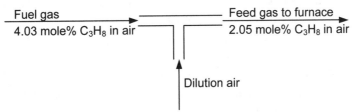

The fuel gas and dilution air would be flowing in separate pipes and would come together at a *tee fitting* in the line, and the combined stream would emerge from the fitting into another pipe leading to the furnace. Not shown in the diagram are valves and flowmeters that would regulate the flow rates of the two inlet streams and a gas sampling tap that would enable the composition of the outlet stream to be monitored.

The process being analyzed in this problem is the mixing of two gases, and the "process unit" is the tee in the piping system where the mixing occurs. To draw and label the process flowchart, we will use a box to represent the tee.

FIGURE 4.11-1: MODEL 2361-FMT GAS/AIR MIXING TEE FROM PYRONICS, INC., CLEVELAND, OHIO.

Name: _____

Date: _____

The flowchart with known amounts and composition variables appears as follows.

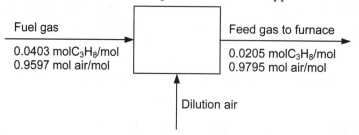

We are told that the fuel-air mixture contains 4.03 mole% of C_3H_8, and since the only other component of the mixture is air, the mixture must also contain 95.97 mole% air. The same reasoning is used to determine the composition of the feed to the furnace, which we are told contains propane at the LFL. The air in all three streams could be broken down into its constituents (oxygen, nitrogen, and other species) in the labeling, but the problem doesn't call for us to do anything that requires it (such as determining the percentage of oxygen in any of the streams) and so for simplicity we treat air as a single species. Notice that we labeled mole fractions rather than percentages on the flowchart and included the units.

The question now is, is the flowchart completely labeled? We suggest taking a moment to look back at the definition of complete labeling and try to answer that question for yourself before proceeding.

The answer is, no. For labeling to be complete, we should be able to express the amount or flow rate of every component of every stream in terms of what is written on the flowchart. We cannot do so for any of the three streams in the process—all we have labeled up to now are component mole fractions. To complete the labeling, we need to define variables for the flow rates (mol/s) of the three streams. Here is the final result.

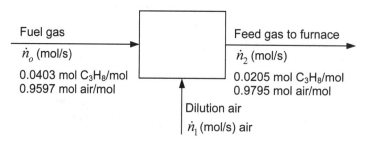

The labeling is now complete. For example, the flow rate of propane in the fuel gas stream can be expressed in terms of labeled quantities as $0.0403\,\dot{n}_0$ (mol C_3H_8/s) [more completely, $(\dot{n}_0\,\text{mol/s})\times$ $(0.0403\,\text{mol }C_3H_8/\text{mol})$], and the same can be done for the air in that stream, the dilution air, and the propane and air in the outlet stream.

(4.11-1)

> Q: Demonstrate that the air feed stream and the outlet gas stream are completely labeled.
> A: Once again, a stream is completely labeled if the flow rate of each of its components can be expressed in terms of labeled constants and variables.
>
> **Air feed (1 component):** Flow rate of air = _____ mol air/s
>
> **Outlet gas (2 components):** Flow rate of propane = _____ mol C_3H_8/s
>
> Flow rate of air = _____ mol air/s

2. Perform a Degree-of-Freedom Analysis.

There are three unknowns on the flowchart ($\dot{n}_0$, $\dot{n}_1$, and $\dot{n}_2$), and since there are two independent species in this nonreactive process (propane and air), we are allowed to write two independent material balances.

DEGREE-OF-FREEDOM ANALYSIS		
UNKNOWNS AND INFORMATION		**JUSTIFICATION/CONCLUSION**
+3 unknowns	$\dot{n}_0, \dot{n}_1, \dot{n}_2$	
−2 balances	propane, air	Two independent species
−0 other info		None given in problem statement
1 DOF		All unknowns cannot be determined

If in Part (a) we had been asked to calculate anything, such as the air-to-fuel mole ratio or the moles of outlet gas per mole of fuel gas, we could stop right here—you can't solve two equations in three unknowns. The benefit of the DOF analysis is that it lets you know this up front, as opposed to your wasting time on an impossible calculation.

We failed to do something, however—namely, choose a basis of calculation. If we specified, for example, 100 mol/s of the fuel gas and replaced $\dot{n}_0$ with that quantity on the flowchart, the DOF analysis would show zero degrees of freedom and the other two variables could be calculated. The calculation would be identical to that in Part (b) (in which a basis is given), so let's go directly there.

(b) If propane flows at a rate of 150 mol C_3H_8/s in the original fuel-air mixture, what is the minimum molar flow rate of the dilution air.

Solution

We are now given a basis of calculation—namely, the flow rate of propane in the fuel-gas mixture. The total molar flow rate of the fuel gas may now be calculated as (fill in the blanks)

$$\dot{n}_0 = \frac{150 \text{ mol } C_3H_8}{s} \bigg| \underline{\hspace{2cm}} = \underline{\hspace{1cm}} \frac{\text{mol fuel gas}}{s} \qquad (4.11\text{-}2)$$

The flowchart and DOF analysis then become

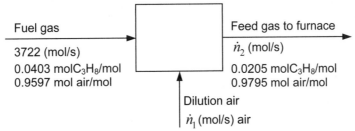

DEGREE-OF-FREEDOM ANALYSIS		
UNKNOWNS AND INFORMATION		**JUSTIFICATION/CONCLUSION**
+2 unknowns	$\dot{n}_1, \dot{n}_2$	—
−2 balances	propane, air	—
0 DOF		Problem is solvable

The problem asked us to calculate the minimum flow rate of dilution air. This would be $\dot{n}_1$, which would drive the concentration of propane in the stream exiting the mixing tee (i.e., the feed to the furnace) down to the lower flammability limit (2.05 mole%). Any lower feed rate of air would allow the concentration in the furnace feed to rise above the LFL and into a danger zone if the furnace is malfunctioning. We can now proceed to the solution.

3. **Write the system equations and circle the variable(s) that will be determined when each equation has been solved.**

Three material balance equations can be written—one on propane, one on air, and one on total moles—but since there are only two independent species *only two of those equations are independent.* (Once propane and air are balanced, for example, the total is automatically balanced.) If we plan to use an equation-solving program such as E-Z Solve, we could choose any two balances and write them in any order, but let's say we plan to solve the problem manually. To determine which balances to write and the order in which to write them, consider how many unknowns each one involves:

(4.11-3)

Total mole balance: 2 unknowns ($\dot{n}_1$ and $\dot{n}_2$) [The balance would be $3722 + \dot{n}_1 = \dot{n}_2$]

Propane balance: _____ unknown(s) (_____),

Air balance: _____ unknown(s) (_____)

Strategy: Write a _____ balance and solve for _____

Then write a _____ balance and solve for _____

For this steady-state nonreactive process, all balances have the form *input = output.* The input and output terms can be read directly from the flowchart, since the chart is completely labeled.

Propane balance: $\left(3722\dfrac{\text{mol}}{\text{s}} \right)\left(0.0403\dfrac{\text{mol C}_3\text{H}_8}{\text{mol}} \right) = \widehat{\dot{n}_2}\left(\dfrac{\text{mol}}{\text{s}} \right)\left(0.0205\dfrac{\text{mol C}_3\text{H}_8}{\text{mol}} \right)$

Total mole balance: _____   (4.11-4)

Since the task was to calculate $\dot{n}_1$, the problem is essentially solved—all that remains is the algebra.

4. **Solve the equations.**

Propane balance $\Rightarrow \dot{n}_2 = 7317$ mol/s

Eq. (4.11-4) $\Rightarrow \dot{n}_1 =$ _____ mol air/s $\xrightarrow{\text{sig. figs.}} \dot{n}_1 = 3600$ mol air/s

(4.11-5)

5. Determine any additional desired quantities.

There are none in this problem but let's make one up as an illustration. Suppose we were asked to determine the molar ratio of dilution air to fuel gas required to achieve the LFL in the feed gas to the furnace. Let's call this ratio R (mol air/mol fuel). Since we have calculated all unknowns on the flowchart, we may then determine any quantity involving those variables: in this case,

$$R = \frac{\underline{\hspace{1cm}} \text{ mol air/s}}{\underline{\hspace{1cm}} \text{ mol fuel/s}} = \underline{\hspace{1cm}} \text{mol air/mol fuel}$$ (4.11-6)

(c) How would the actual dilution air feed rate probably compare with the value calculated in Part (b)? ($>, <, =$)? Explain.

Solution

Suppose you were working in the plant next to that combustion furnace when something went wrong. The air valve opens and just enough air to get to the LFL (based on the expected fuel flow rate) starts going into the furnace. Unfortunately, what went wrong was a malfunction with the fuel gas control valve and the feed gas is going in at a rate 25% higher than it is supposed to be. Then to make things worse, friction in a control valve causes a spark in the feed gas line. What happens next could seriously ruin your day.

Go to the Internet and do a search using the terms "lower flammability limit" + "safety factor", being sure to put both phrases in quotes. What safety factor recommendations do you find for handling flammable gas mixtures? What does this tell you about the flow rate of dilution air that would probably be used on the combustion furnace? Explain below.

(4.11-7)

```
┌─────────────────────────────────────────────────────────┐
│                                                           │
│                                                           │
│                                                           │
│                                                           │
│                                                           │
└─────────────────────────────────────────────────────────┘
```

As we said we would do at the start, we've taken a lot of space to work this problem because of all the explanations, but the solution itself is quite brief. Ideally, your solutions for most problems should be as compact as possible (as long as you provide enough information for them to be comprehensible). For Problem 4.11, your solution should look something like ours shown below (we've left some answers blank). Be sure your balances are dimensionally homogeneous, watch significant figures, include units on all answers, use consistent nomenclature, and write legibly.

Name: _____

Date: _____

VCU CHEMICAL ENGINEERING

PROBLEM 4.11

a)

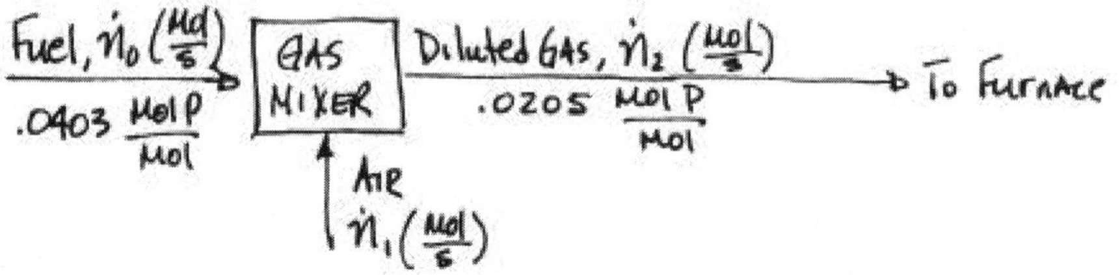

Fuel, $\dot{n}_0 \left(\frac{Mol}{s}\right)$ [GAS MIXER] Diluted GAS, $\dot{n}_2 \left(\frac{Mol}{s}\right)$ → To Furnace

.0403 $\frac{Mol\,P}{Mol}$ .0205 $\frac{Mol\,P}{Mol}$

Air
$\dot{n}_1 \left(\frac{Mol}{s}\right)$

DOFA

3 unknowns ($\dot{n}_0, \dot{n}_1, \dot{n}_2$)
- 2 balances (propane, air)
- 0 info (nothing else given)
_____
 1 DF (need a basis to solve)

b) For 150 $\frac{Mol\,P}{s}$ in fuel inlet, find $\dot{n}_0$ $\frac{Mol\,air}{s}$ MINIMUM

Find $\dot{n}_0$, use as the basis

→ use as basis

$\dot{n}_0 = 150 \frac{Mol\,P}{s} \Big| \underline{\qquad} = \underline{\qquad} \frac{Mol\,Mix}{s}$

3722 $\frac{Mol}{s}$ → [   ] $\dot{n}_2 \left(\frac{Mol}{s}\right)$

.0403 $\frac{Mol\,P}{Mol}$ .0205 $\frac{Mol\,P}{Mol}$

$\dot{n}_1 \left(\frac{Mol}{s}\right)$ air

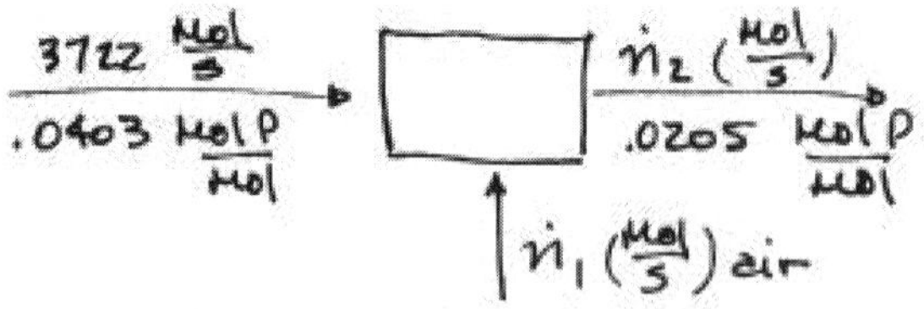

Propane balance: 150 $\frac{Mol\,P}{s}$ = _____ ⟹ $\dot{n}_2$ = ____ $\frac{Mol}{s}$ $\overset{s.f.}{\Rightarrow}$ 7300 $\frac{Mol}{s}$

Total Mols: 3722 $\frac{Mol}{s}$ + $\dot{n}_1$ = $\dot{n}_2$ ⟹ $\dot{n}_1$ = ____ $\frac{Mol}{s}$ $\overset{s.f.}{\Rightarrow}$ 3600 $\frac{Mol}{s}$

c) Better make ___ >> 3600 $\frac{Mol}{s}$ to be sure mixture is below the LEL.

GSH

FIGURE 4.11-2

PROBLEM 4.16

Two aqueous sulfuric acid solutions containing 20.0 wt% H_2SO_4 (SG = 1.139) and 60.0 wt% H_2SO_4 (SG = 1.498) are mixed to form a 4.00 molar solution (SG = 1.213).

(a) Calculate the mass fraction of sulfuric acid in the product solution.

Strategy

By definition, a 4-molar sulfuric acid solution contains 4 g-moles of H_2SO_4 in each liter of solution. To carry out the requested calculation, we need to know how many kg of H_2SO_4 there are in 4.00 mol of H_2SO_4 (the molecular weight tells us this) and how many kg of product solution there are in a liter of product solution (the specific gravity of the product solution gives us this information).

Solution

Start with 4.00 mol H_2SO_4/L solution and write a single dimensional equation to find the mass fraction of H_2SO_4. (Fill in the conversion factors and verify the solution.)

$$\frac{4.00 \text{ mol } H_2SO_4}{\text{L soln.}} \left| \quad \right| \quad \left| \quad \right| = 0.323 \frac{\text{kg } H_2SO_4}{\text{kg soln.}} \qquad \text{(4.16-1)}$$

(b) Taking 100 kg of the 20% feed solution as a basis, draw and label a flowchart of this process, labeling both masses and volumes, and do the degree-of-freedom analysis. Calculate the feed ratio (liters 20% solution/liter 60% solution).

Solution

1. Draw the flowchart, choose a basis of calculation (already done), and label the flowchart.

We'll get you started on the flowchart and you fill in the rest.

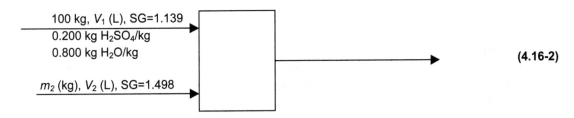

$$\text{(4.16-2)}$$

After you finish labeling the chart, before you go any further verify that the chart is completely labeled—that you can express the masses of H_2SO_4 and H_2O in terms of labeled quantities.

2. Perform a Degree-of-Freedom Analysis.

Some information is already filled in; you fill in the remainder.

$$\text{(4.16-3)}$$

DEGREE-OF-FREEDOM ANALYSIS		
UNKNOWNS AND INFORMATION		**JUSTIFICATION/CONCLUSION**
+ <u> 5 </u> unknowns		
– ___ balances		___ independent species
– ___ SGs (densities)	ρ_1, ρ_2, ρ_3	Relate masses to volumes
= 0 DOF		Problem is solvable

Name: _____

Date: _____

3. **Write the system equations in an efficient order (if there is one) and circle the unknown(s) to be found by solving each equation.**

Be sure to label each balance written with the component being balanced. If you have to solve any equations simultaneously, enclose the equations in brackets and circle the unknowns to be determined. Don't do any algebra or arithmetic yet.

Equations for Part (b). (The first two are given)   (4.16-4)

Density of 20% solution : $\left(V_1\right) = \dfrac{100 \text{ kg}}{} \left|\dfrac{\text{L}}{1.139 \text{ kg}}\right.$

Total mass balance : $100 \text{ kg} + \left(m_2\right) = \left(m_3\right)$

4. **Solve the equations.**

Complete the solution to Part (b) by solving the equations (consider using E-Z Solve to make life easier for yourself), report the values for the unknowns in the spaces provided below (you should have five unknowns).

(4.16-5)

_____ =

_____ =

_____ =   (UNITS!)

_____ =

_____ =

Feed ratio = _____ L 20% solution/L 60% solution   (4.16-6)

5. **Determine any remaining quantities requested in the problem statement.**

Part (c) requests a feed rate calculation so we'll finish with that.

(c) What feed rate of the 60% solution (L/h) would be required to produce 1250 kg/h of the product?

Strategy

This is a scaleup calculation. From your answers above, you already know the volume of 60% solution (the calculated value of V_2) needed to produce a known mass of product solution (the calculated value of m_3). The requested volumetric feed rate in L/h is determined by multiplying V_2 (L) by the *scaling factor* [1250 kg/h]/[m_3 (kg)]. Another way of setting up the calculation is in terms of ratios:

$$\frac{V_{reqd} \text{ (L/h)}}{V_2 \text{ (L)}} = \frac{1250 \text{ kg/h}}{m_3 \text{ (kg)}}$$

Solution

Calculate the feed rate of the 60% solution using a single dimensional equation. It is important to write out the units *completely*. Writing "1250 kg/h" to start is insufficient—we need to write "1250 kg product solution/h" to specify the quantity correctly. Write the units completely every time and you will rarely make a mistake on this kind of calculation.

(4.16-7)

$$\frac{1250 \text{ kg product solution}}{h} \left| \frac{\_\_\_\_\_ \text{ L 60\% solution}}{\_\_\_\_\_ \text{ kg product solution}} \right| = \frac{\_\_\_\_\_ \text{ L 60\% solution}}{h}$$

Notes and Calculations

PROBLEM 4.21

A dilute aqueous solution of H_2SO_4 (Solution A) is to be mixed with a solution containing 90.0 wt% H_2SO_4 (Solution B) to produce a 75 wt% solution (Solution C).

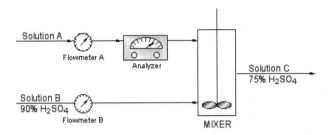

The flow rate and concentration of Solution A change periodically, so that it is necessary to adjust the flow rate of Solution B to keep the product H_2SO_4 concentration constant.

Flowmeters A and B have linear calibration plots of mass flow rate ($\dot{m}$) versus meter reading (R), which pass through the following points:

$$\text{Flowmeter A:} \quad \dot{m}_A = 150 \text{ lb}_m/\text{h, } R_A = 25$$
$$\dot{m}_A = 500 \text{ lb}_m/\text{h, } R_A = 70$$
$$\text{Flowmeter B:} \quad \dot{m}_B = 200 \text{ lb}_m/\text{h, } R_B = 20$$
$$\dot{m}_B = 800 \text{ lb}_m/\text{h, } R_B = 60$$

The analyzer calibration is a straight line on a semilog plot of %H_2SO_4 (x) on a logarithmic scale versus meter reading (R_x) on a linear scale. The line passes through the points ($x = 20\%$, $R_x = 4.0$) and ($x = 100\%$, $R_x = 10.0$).

(a) Calculate the flow rate of Solution B needed to process 300 lb$_m$/h of 55.0% H_2SO_4 (Solution A), and the resulting flow rate of Solution C. (The calibration data are not needed for this part.)

Strategy

Part (a) is a mixing problem. We need to sketch a flowchart and label it completely, then do the degree-of-freedom analysis and determine if we have enough information to calculate the requested flow rates. If we do, we'll complete the solution. Follow the steps in sequence.

Solution

1. Draw the flowchart, choose a basis of calculation, and complete the flowchart labeling.

The basis of calculation for this problem is given in the problem statement. Report it below.

Basis: _____ (UNITS!)   **(4.21-1)**

We've drawn the mixer and streams; you finish labeling the diagram. Use $\dot{m}_B$ and $\dot{m}_C$ (lb$_m$/h) for the flow rates of Solutions B and C, respectively. Remember to label the stream compositions as mass fractions with appropriate units.

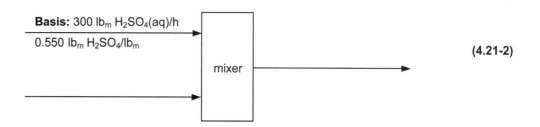

Basis: 300 lb$_m$ H$_2$SO$_4$(aq)/h

0.550 lb$_m$ H$_2$SO$_4$/lb$_m$

mixer

(4.21-2)

2. Perform a Degree-of-Freedom Analysis.

Fill in the table below. (Refer back to the previous problem solution if you need a hint.) There are zero degrees-of-freedom for this flowchart.

(4.21-3)

DEGREE-OF-FREEDOM ANALYSIS		
UNKNOWNS AND INFORMATION		**JUSTIFICATION/CONCLUSION**
+2 unknowns	(____ , ____)	
0 DOF		Problem is solvable

3. Write the system equations in an efficient order (if there is one) and circle the unknown(s) to be found by solving each equation. Be sure to label each balance written with the component being balanced.

(1) **Total mass balance:** _____

(4.21-4)

(2) _____ : _____

(4.21-5)

4. Solve the equations. Use E-Z Solve. Print the solution table and tape a copy of the solution in the space below. Write the values obtained for the flow rates in the spaces provided for them.

(4.21-6)

$\dot{m}_B =$ _____

(UNITS!)

$\dot{m}_C =$ _____

5. Determine any remaining quantities requested in the problem statement.

Are there any?   ❒ Yes   ❒ No

(b) Derive the calibration equations for $\dot{m}_A(R_A), \dot{m}_B(R_B),$ and $x(R_x)$. Calculate the values of R_A, R_B, and R_x corresponding to the flow rates and concentrations of Part (a).

Strategy

Two calibration points are given for each measurement device. We will write the equation appropriate to each calibration plot, determine a straight line form for each equation, use the two calibration points given to calculate the slope and intercept of each linear equation, and, when appropriate, rewrite the calibration equation in its nonlinear form. The calibration equations are linear for Flowmeters A and B so no rewriting will be needed in those cases.

Solution

We'll walk you through the calculations for Flowmeter A and the analyzer, and you complete the calculations and do the one for Flowmeter B.

Flowmeter A: The equation for Flowmeter A is a straight line: $\dot{m}_A = a \cdot R_A + b$, where a and b are the slope and intercept, respectively. We write this equation for each of the calibration points supplied and subtract one equation from the other to eliminate the intercept, b. The slope can then be determined from the resulting equation. The slope and a calibration point ($R=25$, $\dot{m}=150$ lb$_m$/h) are then used to determine the intercept.

$$\dot{m}_A = a \cdot R_A + b \quad , (\dot{m}_{A1} = 150 \text{ lb}_m/\text{h}, R_{A1} = 25), (\dot{m}_{A2} = 500 \text{ lb}_m/\text{h}, R_{A2} = 70)$$

$$a = \frac{\dot{m}_{A2} - \dot{m}_{A1}}{R_{A2} - R_{A1}} = \frac{(500 - 150) \text{ lb}_m/\text{h}}{70 - 25} = 7.778 \text{ lb}_m/\text{h}$$

$$150 \frac{\text{lb}_m}{\text{h}} = \left(7.778 \frac{\text{lb}_m}{\text{h}} \right) \cdot 25 + b \Rightarrow b = -44.44 \frac{\text{lb}_m}{\text{hr}}$$

$$\Rightarrow \boxed{\dot{m}_A = 7.778 R_A - 44.444} \quad \text{Flowmeter A Equation}$$

Note: When doing curve fitting, it is advisable to retain extra significant figures for the slope and intercept to keep from losing precision, and to round off to the correct number of significant figures when the derived equation is used to calculate a system variable.

Flowmeter B: Your turn. Show your work and write the final equation in the small box.

(4.21-7)

$$\dot{m}_B = a \cdot R_B + b \quad , (\dot{m}_{B1} = 200 \text{ lb}_m/\text{h}, R_{B1} = 20), (\dot{m}_{B2} = 800 \text{ lb}_m/\text{h}, R_{B2} = 60)$$

Flowmeter B Equation

H_2SO_4 Analyzer: This one requires a bit more work but once we get the equation into linear form, the solution proceeds as before. We are told that the analyzer calibration is a straight line on a semi-log plot of x versus R_x. Therefore, the equation for the analyzer must be the exponential equation: $x = c \cdot e^{aR_x}$. For this equation to be dimensionally homogeneous, the constant c must have the same units as x. The product aR_x must be dimensionless since it is the argument of the exponent, and since R_x is just a number on a numerical scale (no units), a must also be dimensionless. The calculation proceeds as shown in Section 2.7d of the text.

$$x\% = ce^{aR_x} \Rightarrow \ln x = aR_x + \ln c \ , \quad (x_1 = 20\%, R_{x1}{=}4.0), \ (x_2 = 100\%, R_{x2}{=}10.0)$$

$$a = \frac{\ln(\underline{\quad} / \underline{\quad})}{\underline{\quad} - \underline{\quad}} = 0.2682 \tag{4.21-8}$$

$$\ln c = \underline{\quad}$$

$$\Rightarrow c = \underline{\qquad} = 6.84\%$$

Write the complete analyzer equation in exponential form in the box below.

$$\boxed{} \tag{4.21-9}$$

H$_2$SO$_4$ Analyzer Equation

(c) The process technician's job is to read Flowmeter A and the analyzer periodically, and then to adjust the flow rate of Solution B to its required value. Derive a formula that the technician can use for R_B in terms of R_A and R_x, and then check it by substituting the values of Part (a).

Strategy

Very often, when faced with a "derive" task, students will freeze and declare "I don't know where to start!" Here is a simple way to derive an expression relating specified system variables (in this problem, R_A, R_B, and R_x). The first step is to write all known relationships between the specified variables and other system variables. In this problem, we know the relationships between the flowmeter readings (R_A and R_B) and the flow rates of the respective streams they measure ($\dot{m}_A$ and $\dot{m}_B$), and between the analyzer reading (R_x) and the percentage of sulfuric acid in Stream A (x). The next step is to write all known equations relating the other system variables ($\dot{m}_A, \dot{m}_B, \dot{m}_C,$ and x), which in our system would be two material balances. Now it's just an algebra problem, solving equations for unspecified variables (all but R_A, R_B, and R_x), substituting for those variables in other equations, until we are left with the desired relationship.

Solution

Let's write the calibration equations and material balances and see what we've got.

Flowmeter A calibration : $\dot{m}_A = 7.778R_A - 44.444$ (4.21-10)

Flowmeter B calibration : $\dot{m}_B = 15.0R_B - 100$ (4.21-11)

Analyzer calibration : $x\% = 6.84e^{0.2682R_x}$ (4.21-12)

Total mass balance : _____ (4.21-13)

H$_2$SO$_4$ balance : _____ (4.21-14)

The strategy is now to substitute the expression for $\dot{m}_C$ from Eq. (4.21-13) into Eq. (4.21-14), and then substitute into the resulting equation the expressions for $\dot{m}_A, \dot{m}_B,$ and x of Eqs. (4.21-10), (4.21-11), and (4.21-12) to obtain an expression relating R_A, R_B, and R_x.. (Don't worry if the result looks ugly, and don't simplify it yet.)

(4.21-15)

Our final task will be to solve this expression for R_B to obtain the requested formula—which, presumably, will be possible since the problem asked for it. Your answer should begin with the expression $R_B = \left(2.59 - 0.236e^{0.2682 \cdot R_x}\right)R_A + \ldots.$

Finally, for $R_A = 44.3$ and $R_x = 7.78$, prove your equation yields $R_B = 33.3$. Substitute this value for R_B into (4.21-11), calculate $\dot{m}_B$, and check your answer against the one you got in Part (a).

Name: _____

Date: _____

Notes and Calculations

PROBLEM 4.32

Fresh orange juice contains 12.0 wt% solids and the balance water, and concentrated orange juice contains 42.0 wt% solids. Initially, a single evaporation process was used for the concentration, but volatile constituents of the juice escaped with the water, leaving the concentrate with a flat taste. The current process overcomes this problem by bypassing the evaporator with a fraction of the fresh juice. The juice that enters the evaporator is concentrated to 58.0 wt% solids, and the evaporator product stream is mixed with the bypassed fresh juice to achieve the desired final concentration.

(a) Draw and label a flowchart of this process, neglecting the vaporization of everything in the juice but water. First prove that the subsystem containing the point where the bypass stream splits off from the evaporator feed has one degree of freedom. (If you think it has zero degrees, try determining the unknown variables associated with this system.) Then perform the degree-of-freedom analysis for the overall system, the evaporator, and the bypass-evaporator product mixing point, and write in order the equations you would solve to determine all the unknown stream variables. In each equation, circle the variable for which you would solve, but don't do any calculations.

Strategy

Read the problem statement again. We are given several mass fractions in the form of percentages, but no basis for the flowchart until we get to Part (b). We'll use the basis given there, 100 kg of fresh juice as the feed to the process. This process has a bypass stream around the evaporator and a mixing point where the bypass stream rejoins the product stream from the evaporator. Bypass splitting points and stream mixing points are flowchart elements about which balances may be written, as are overall systems (that enclose all recycle and bypass streams) and individual process units.

Flowchart elements can be combined to form *subsystems*—parts of the flowchart for which the system equations are solvable. Often, solving the equations for a subsystem supplies the information needed to make another subsystem solvable. Our strategy will be to locate all of the subsystems on the flowchart and then determine, using degree-of-freedom analyses, whether the balances on the overall system or one or more of the subsystems can be solved. If so, we'll write the equations needed to determine missing variables for that subsystem, then move to another solvable subsystem, and in this manner work through as many subsystems as necessary to complete the solution of the problem.

Solution

We use boxes with dashed line borders to enclose subsystems. The flowchart below contains a partially labeled flowchart with the overall system boundary and the boundary of one subsystem (the bypass splitting point) shown explicitly. You should complete the labeling of the streams and draw the two other simple subsystem boundaries. For more guidance, study how the subsystems are shown in Example 4.7-2 on p. 135 in the text.

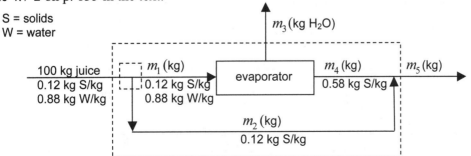

S = solids
W = water

m_3 (kg H$_2$O)

100 kg juice
0.12 kg S/kg
0.88 kg W/kg

m_1 (kg)
0.12 kg S/kg
0.88 kg W/kg

evaporator

m_4 (kg)
0.58 kg S/kg

m_5 (kg)

m_2 (kg)
0.12 kg S/kg

Following is the DOF analysis for the bypass splitting point and the mass balance equation for this element. Remember, the only variables you count are those associated with the streams that intersect the boundary of the subsystem you are analyzing. (Variables associated with other streams will not enter the balance equations for this one.) Verify that this is the case in the analysis that follows.

DEGREE-OF-FREEDOM ANALYSIS: BYPASS SPLIT		
UNKNOWNS AND INFORMATION		**JUSTIFICATION/CONCLUSION**
+2 unknowns	(m_1, m_2)	
−1 balance		
1 DOF		Not immediately solvable.

Total mass balance on the bypass split: $100 = m_1 + m_2$

There are two components in the bypass split subsystem; S and H_2O. Why can only one balance be written? Answer below.   **(4.32-1)**

To convince yourself of your answer (or to get a hint if you didn't get the answer), try writing a sugar balance on the bypass split and solving it simultaneously with the total mass balance. Explain the result you obtain.

Now complete DOF analyses for the overall system and the two other subsystems you drew on the chart.   **(4.32-2)**

DEGREE-OF-FREEDOM ANALYSIS: OVERALL SYSTEM		
UNKNOWNS AND INFORMATION		**JUSTIFICATION/CONCLUSION**
+2 unknowns	(___ , ___)	
−2 balances		
0 DOF		System equations are solvable

Let's assume that we will begin the solution by writing balances on the overall system and solving for the two unknown variables associated with it (m_3 and m_5), which we have just proved we can do. When these variables appear in other subsystem DOF analyses, they should no longer be counted as unknowns. Let's try the bypass-product stream mixing point next.   **(4.32-3)**

DEGREE-OF-FREEDOM ANALYSIS: MIXING POINT		
UNKNOWNS AND INFORMATION		**JUSTIFICATION/CONCLUSION**
+ ___ unknowns	(___ , ___)	
− ___ balances		
0 DOF		System equations are solvable

The only remaining unknown on the chart now is m_1. We can now go back to the bypass splitting point to find it.   **(4.32-4)**

DEGREE-OF-FREEDOM ANALYSIS: BYPASS SPLIT		
UNKNOWNS AND INFORMATION		**JUSTIFICATION/CONCLUSION**
+1 unknown	(___)	
−1 balance		
0 DOF		System equation is solvable

At this point, all of the initially unknown variables on the chart may be considered determined. Balances could still be written on the evaporator to provide a consistency check of the calculated values, but the exercise would not yield new information.

We can now see the power of degree-of-freedom analysis. Using it, we can proceed to lay out an efficient solution algorithm for the problem, quickly writing just the equations we need and then solving them. What students normally do with bypass and recycle problems is just plunge in and start writing equations, hoping something good will happen. It may, but when the systems start getting more complex it may be hours before it does. It also may never happen—it is not uncommon for students to spend four or five hours on a complex recycle problem and end up with nothing but a jumble of unsolved equations. Contrast that with what follows. We'll circle variables for which the equations will be solved, enclosing simultaneous equations in brackets.

Overall system sugar balance : $100(0.12) = 0.42\,\widehat{m_5}$

Overall system mass balance : $100 = \widehat{m_3} + m_5$

Mixing point mass balance : $\left\{ \widehat{m_4} + \widehat{m_2} = m_5 \right.$

Mixing point sugar balance : $\left. 0.58m_4 + 0.12m_2 = 0.42m_5 \right\}$ (m_5 is now known)

Bypass split point mass balance : $100 = \widehat{m_1} + m_2$

All that remains is algebra, or even simpler, solving the equations with E-Z Solve.

(b) Calculate the amount of product (42% concentrate) produced per 100 kg fresh juice fed to the process and the fraction of the feed that bypasses the evaporator.

Use E-Z Solve to solve the equations you wrote in Part (a).

(4.32-5)

$m_1 =$ _____ kg fresh juice fed to evaporator

$m_2 =$ _____ kg fresh juice bypassed

$m_3 =$ _____ kg water evaporated

$m_4 =$ _____ kg 58% concentrate leaving evaporator

$m_5 =$ _____ kg 42% concentrate (product)

Fraction that bypasses $= \dfrac{m_2 \text{ kg bypassed}}{100 \text{ kg fresh feed}} = \dfrac{\text{kg bypassed}}{\text{kg fresh feed}}$

(c) Most of the volatile ingredients that provide the taste of the concentrate are contained in the fresh juice that bypasses the evaporator. You could get more of these ingredients in the final product by evaporating to (say) 90% solids instead of 58%; you could then bypass a greater fraction of the fresh juice and thereby obtain an even better tasting product. Suggest possible drawbacks to this proposal.

(4.32-6)

Notes and Calculations

PROBLEM 4.36

In the production of a bean oil, beans containing 13.0 wt% oil and 87.0% solids are ground and fed to a stirred tank (the *extractor*) along with a recycled stream of liquid n-hexane. The feed ratio is 3 kg hexane/kg beans. The ground beans are suspended in the liquid, and essentially all of the oil in the beans is extracted into the hexane. The extractor effluent passes to a filter. The filter cake contains 75 wt% bean solids and the balance bean oil and hexane, the latter two in the same ratio in which they emerge from the extractor. The filter cake is discarded and the liquid filtrate is fed to a heated evaporator in which the hexane is vaporized and the oil remains as a liquid. The oil is stored in drums and shipped. The hexane vapor is subsequently cooled and condensed, and the liquid hexane condensate is recycled to the extractor.

(a) Draw and label a flowchart of the process, do the degree-of-freedom analysis, and write in an efficient order the equations you would solve to determine all unknown stream variables, circling the variables for which you would solve.

Hx = hexane, S = solids

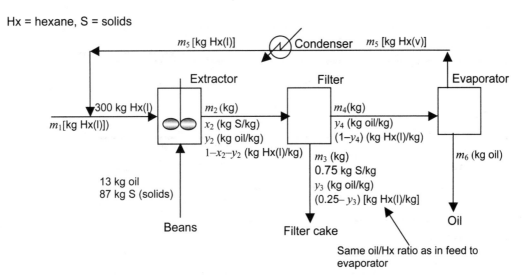

We've taken a basis of 100 kg beans to the extractor. Given the hexane/beans feed ratio (=3) to the extractor, we can immediately specify the quantity of hexane fed to this vessel as well.

Strategy

In this system, there are three process units and a mixing point where the recycle stream joins the feed. Balances can be written around the overall system and each of these subsystems. (Why isn't a balance around the condenser helpful?)

(4.36-1)

Following is an outline of how degree-of-freedom analyses might be used to plan the solution for all unknown system variables. (Fill in the missing information.)

Name: _____

Date: _____

(4.36-2)

DEGREE-OF-FREEDOM ANALYSIS: OVERALL	
UNKNOWNS AND INFORMATION	**JUSTIFICATION/CONCLUSION**
+ 4 unknowns ___, ___, ___, ___	
− 3 balances ___, ___, ___	
1 DOF	Don't start with overall balances

DEGREE-OF-FREEDOM ANALYSIS: EXTRACTOR	
UNKNOWNS AND INFORMATION	**JUSTIFICATION/CONCLUSION**
+ _____ unknowns _____	
− _____ balances _____	
0 DOF	Start here. In subsequent analyses, treat m_2, x_2, and y_2 as known

DEGREE-OF-FREEDOM ANALYSIS: FILTER	
UNKNOWNS AND INFORMATION	**JUSTIFICATION/CONCLUSION**
+ _____ unknowns _____	
− _____ balances _____	
− 1 _____	
0 DOF	Do this system next. m_3, y_3, m_4, and y_4 are then known.

DEGREE-OF-FREEDOM ANALYSIS: EVAPORATOR	
UNKNOWNS AND INFORMATION	**JUSTIFICATION/CONCLUSION**
+ _____ unknowns _____	
− _____ balances _____	
0 DOF	

DEGREE-OF-FREEDOM ANALYSIS: MIXING POINT	
UNKNOWNS AND INFORMATION	**JUSTIFICATION/CONCLUSION**
+ _____ unknowns _____	
− _____ balances _____	
0 DOF	

Solution

Here are the equations with the variables to be determined circled. Fill in the blanks, and notice how the equations are written to minimize the number that must be solved simultaneously.

Mass balance on extractor: $(300 + 87 + 13) \text{ kg} = \boxed{m_2}$

S balance on extractor: $87 \text{ kg S} = m_2 \boxed{x_2}$

(4.36-3)

Oil balance on extractor: _____

Oil/hexane ratio equality (filter cake & filter inlet): $\dfrac{y_3}{0.25 - y_3} = \dfrac{y_2}{1 - x_2 - y_2}$

Mass balance on filter: $m_2 = m_3 + m_4$

S balance on filter: _____

Oil balance on filter: _____ $= m_3 y_3 + m_4 y_4$

⎬ Solve simultaneously

Hx balance on evaporator: _____ $= m_5$

Oil balance on evaporator: _____ $= m_6$

Mass balance on mix point: $m_1 +$ _____ $=$ _____

We now have a complete and efficient set of equations for solving the flowchart. Do the algebra or (much easier) use E-Z Solve to determine the values of the stream variables listed below.

(4.36-4)

$m_2 =$ _____ kg $x_2 =$ _____ kg S/kg $y_2 =$ _____ kg oil/kg

$m_3 =$ _____ kg $y_3 =$ _____ kg oil/kg

$m_4 =$ _____ kg $y_4 =$ _____ kg oil/kg

$m_5 =$ _____ kg

$m_6 =$ _____ kg

$m_1 =$ _____ kg

(b) Calculate the yield of bean oil product (kg oil/kg beans fed), the required fresh hexane feed (kg C_6H_{14}/kg beans fed), and the recycle to fresh feed ratio (kg hexane recycled/kg fresh feed).

Solution

The answer to Part (b) is trivial since all the stream variables are known. Remember that the total bean feed is $87 + 13 = 100$ kg beans. Using the values you determined above, calculate the ratios requested and record them in the following box.

(4.36-5)

These ratios deserve a few words. Virtually all processes are characterized by those who work on them by using a few key parameters—often yields, feed ratios, and recycle ratios. Since we cannot possibly keep all of the various stream flows and mass fractions in our heads, the use of a few key indicators of process stability and productivity are used. In chemical plants that produce hundreds of millions of kg of product per year, fractions of a percent improvement in product yields can result in millions of dollars in increased annual income. Likewise, tiny reductions in the amount of solvent used to process a pound of product or small reductions in the amount of material that must be recycled can have huge environmental and financial consequences due to the shear tonnage of material being handled in the plant. The repeated requests for such calculations in the text are true to form and very realistic. Practice calculating and understanding process ratios; before long, you may be asked to "think" in terms of such key indicators.

(c) It has been suggested that a heat exchanger might be added to the process. This process unit would consist of a bundle of parallel metal tubes contained in an outer shell. The liquid filtrate would pass from the filter through the inside of the tubes and then go on to the evaporator. The hot hexane vapor on its way from the evaporator to the extractor would flow through the shell, passing over the outside of the tubes and heating the filtrate. How might the inclusion of this unit lead to a reduction in the operating cost of the process.

Solution

(4.36-6)

| |
| |
| |
| |

This problem is intended to give you a sense of what a heat exchanger looks like and what its function is in a chemical process. Do yourself a favor—every time you encounter something like this in a problem, go find a picture of the process unit and examples of how it is used. Doing so will enable you to make sense of problems that might otherwise be confusing, and it will also give you a preview of aspects of your working environment after you graduate. A good place to start is the *Visual Encyclopedia of Chemical Engineering Equipment* on the CD that came with your text.

The picture at the right came from Federal Equipment Company's internet site (http://www.fedequip.com). The cap that covers the tube bundle is open so the tube bundle is visible at the lower right of the picture. The ***tube side*** fluid travels through these tubes and exits at the other end of the unit. The big nozzle on the side is where the ***shell side*** fluid either enters or exits—hard to tell which from the picture. In any case, the shell side fluid contacts the tubes and heat is transferred from one fluid to the other across the tube walls. This heat exchanger has about 1300 ft^2 of surface area on the tubes for heat transfer.

FIGURE 4.36-1: A SHELL-AND-TUBE HEAT EXCHANGER.

Heat exchangers are all around us. The radiator in your car is a heat exchanger although, internally, it is a bit different from the *shell-and-tube* heat exchanger described in Part (c). A steam iron is a heat exchanger; the heating coils inside transfer heat to water in tubes and turn the water to steam. The coils on the back of your refrigerator are a heat exchanger. These coils release heat from the internal coolant to the air behind the fridge. That's why you are supposed to vacuum all the dust away from behind your refrigerator periodically. If the dust accumulates, it harms the heat transfer from the coils and ruins the efficiency of the unit. The air conditioner or heat pump on your house is a heat exchanger—another radiator-type exchanger similar to the one in your car.

(d) Suggest additional steps that might improve the process economics.

Strategy

This is an excellent exercise for group brainstorming and, again, is a very realistic request. Process engineers spend a lot of hours in group discussions trying to come up with new ideas for process improvements. If possible, get a few friends and spend a few minutes brainstorming this question with a copy of the flowchart on a marker board. See how many ideas you can come up with.

Here are some hints to start you off:
- the oil exits the evaporator hot
- hexane costs more than butane
- the wetcake contains 25 wt% liquid
- salad dressing washes off your salad bowl faster with hot water than it does with cold water
- 300 kg of hexane just to dissolve 13 kg of oil?

Notes and Calculations

PROBLEM 4.50

Ethane is chlorinated in a continuous reactor:

$$C_2H_6 + Cl_2 \longrightarrow C_2H_5Cl + HCl$$

Some of the product monochloroethane is further chlorinated in an undesired side reaction:

$$C_2H_5Cl + Cl_2 \longrightarrow C_2H_4Cl_2 + HCl$$

(a) Suppose your principal objective is to maximize the selectivity of monochloroethane production relative to dichloroethane production. Would you design the reactor for a high or low conversion of ethane? Explain your answer. (*Hint*: If the reactor contents remained in the reactor long enough for most of the ethane in the feed to be consumed, what would the main product constituent probably be?) What additional processing steps would almost certainly be carried out to make the process economically sound?

(4.50-1)

(b) Draw and label a flowchart, assuming that the reactor feed contains only the two reactants. Use a degree-of-freedom analysis based on atomic species balances to determine how many process variable values must be specified for you to be able to calculate the remainder.

The flowchart is shown below. Verify that all species have been accounted for, and then complete the degree-of-freedom analysis.

n_1 (mol C_2H_6) → | Chlorination Reactor | → 100 mol C_2H_5Cl

n_2 (mol Cl_2)

n_3 (mol C_2H_6)
n_4 (mol HCl)
n_5 (mol $C_2H_4Cl_2$)
n_6 (mol Cl_2)

If you had been asked to draw the flowchart yourself, would you have remembered to include unreacted ethane (n_3) and chlorine (n_6)? In commercial chemical reactors, there is almost always some unconsumed reactant in the product gas, because the small amount of additional product that could be obtained by converting it would not justify the large additional reactor volume it would require to do it. Also, in most reactors all but one reactant is fed in excess, so that even if the limiting reactant is completely consumed, there would always be some of the other reactants in the product.

(4.50-2)

DEGREE-OF-FREEDOM ANALYSIS: CHLORINATION REACTOR	
UNKNOWNS AND INFORMATION	JUSTIFICATION/CONCLUSION
+ __ unknowns _____	
− __ balances _____	if using *atomic species balances*
3 DOF _____	

(c) The reactor is designed to yield 15% conversion of ethane and chloroethane to dichloroethane selectivity of 14 mol C_2H_5Cl/mol $C_2H_4Cl_2$, with a negligible amount of chlorine in the product gas. Calculate the feed ratio (mol Cl_2/mol C_2H_6) and the fractional yield of monochloroethane (moles actually produced/theoretical maximum).

Strategy

We have been given three new items of information about the process: (1) conversion is 15%, (2) selectivity is 14, and (3) there is no Cl_2 in the product stream. Since there were three degrees of freedom, we should now be able to solve for all unknown variables on the flowchart.

Generally, it's a good idea to translate what we know about the process into equations before starting to write material balances. We'll do so with the three items just listed, solve for as many variables as we can, and then write the balances to complete the solution.

Solution

No Cl_2 remains in the product stream $\Rightarrow$ _____ = 0   (4.50-3)

Selectivity = 14 (mol C_2H_5Cl/mol $C_2H_4Cl_2$ in product)

$$\Rightarrow \frac{\underline{\quad} \ \text{mol } C_2H_5Cl}{\underline{\quad} \ (\text{mol } C_2H_4Cl_2)} = 14 \quad \Rightarrow \quad n_5 = \underline{\quad} \ \text{mol } C_2H_5Cl_2 \qquad (4.50\text{-}4)$$

Ethane Conversion = 15%. If 15% of the ethane fed to the reactor ($= n_1$) reacts, 85% remains unreacted so 85% of the ethane fed must appear in the product stream exiting the reactor ($= n_3$). Thus,

$$\underline{\quad} = 0.85 \times \underline{\quad} \qquad (4.50\text{-}5)$$

We were not able to solve for a variable but at least we have an equation relating two of the four remaining unknowns. Once we write our three allowed atomic species balances we will have four equations in the four unknowns, and all that will remain is the algebra.

We first write the atomic carbon balance. (Verify the balance by inspecting the flowchart.)

Carbon balance

$$\underline{\quad} \frac{(\text{mol } C_2H_6)}{} \left| \frac{\underline{\quad} \ \text{mol C}}{\text{mol } C_2H_6} = \underline{\quad} \frac{\text{mol } C_2H_5Cl}{} \right| \frac{\underline{\quad} \ \text{mol C}}{\text{mol } C_2H_5Cl}$$

$$+ \ \underline{\quad} \frac{(\text{mol } C_2H_6)}{} \left| \frac{\underline{\quad} \ \text{mol C}}{\text{mol } C_2H_6} + \frac{n_5 \ (\text{mol } C_2H_4Cl_2)}{} \right| \frac{\underline{\quad} \ \text{mol C}}{\text{mol } C_2H_4Cl_2} \qquad (4.50\text{-}6)$$

$$\Rightarrow \ n_1 = 100 + n_3 + n_5$$

Write the remaining two atomic species balances, remembering that there is no Cl_2 in the product.

Cl balance _____   (4.50-7)

H balance _____   (4.50-8)

Now, look at the unknowns in the equations. The carbon balance contains n_1 and n_3, as does the equation derived from the ethane conversion specification. Once we have solved those two equations simultaneously for n_1 and n_3, we can find n_4 from the hydrogen balance and then n_2 from the chlorine balance.

It is not always easy to simply write down the solution equations for problems in an efficient order, especially as the problems get more complex. If you plan to solve equations by hand, get a piece of scrap paper and write all of the equations you are allowed to write. Then, look at them and identify those that allow you to isolate and calculate a single unknown. If you cannot find any equations that allow this, look next for two equations in two unknowns and so on until you are able to find the best possible approach for solving the problem. As you work more problems, you will get better at quickly finding efficient solutions. *Alternatively, simply enter the equations in E-Z Solve in any order and let the program solve them.* The more equations you have, the more time E-Z Solve can save you, especially if some of the equations are nonlinear (which is not the case in this problem).

The equations, written in the most efficient order, are

$$\boxed{n_5} = 100/14$$

$$\left.\begin{array}{l}(1-0.15)\boxed{n_1}=\boxed{n_3} \\ n_1 = 100 + n_3 + \left(\dfrac{100}{14}\right)\end{array}\right\} \text{ Solve simultaneously for } n_1 \text{ and } n_3.$$

$$6n_1 = 5(100) + 6n_3 + \boxed{n_4} + 4n_5$$

$$2\boxed{n_2} = 100 + n_4 + 2n_5$$

Let's let E-Z Solve solve the equations. The code, including comments and documentation (it's very important to include both), is:

E-Z Solve program

```
//Problem 4.50 - Chlorination of ethane to chloroethane and dichloroethane
/*
Variables and units
n1 = ethane in feed, mol; n2 = chlorine in feed, mol; n3 = unreacted ethane in product, mol
n4 = HCl in product, mol; n5 = dichloroethane in product, mol
*/
n5 = 100/14                          //selectivity specification
0.85*n1 = n3                         //conversion specification
n1 = 100 + n3 + n5                   //carbon mol balance
6*n1 = 5*(100) + 6*n3 + n4 + 4*n5    //hydrogen mol balance
2*n2 = 100 + n4 + 2*n5               //chlorine mol balance
```

Enter the code into E-Z Solve and compute the values of all of the variables. Write the values reported in E-Z Solve's output table in the box. (Watch significant figures!)

(4.50-9)

$n_1 = $ _____	$n_2 = $ _____	$n_3 = $ _____
$n_4 = $ _____	$n_5 = $ _____	$n_6 = 0$ (given)

Name: _____

Date: _____

We can now compute the feed ratio requested (mol Cl_2/mol C_2H_6) and the fractional yield. The feed ratio is easy—use the values obtained from E-Z Solve to compute it (or you could add another equation to your code and let E-Z Solve do it for you). Compute the value and record it below.

Feed ratio: mol Cl_2/mol C_2H_6 = _____   **(4.50-10)**

To compute the fractional yield, we need to know the maximum amount of the desired product that could be produced, which is the amount that would be formed if <u>all</u> of the limiting reactant is converted to the desired product. From the stoichiometric equation for the formation of chloroethane, one mole of Cl_2 is required for each mole of C_2H_6. Since the feed ratio of ethane to chlorine is greater than 1, Cl_2 is the limiting reactant. (We could also have known this from the fact that there is no chlorine left in the product.) Therefore, the maximum possible moles of C_2H_5Cl formed is:

$$(n_{C_2H_5Cl})_{max} = \underline{\qquad} \frac{\text{mol } Cl_2 \text{ react}}{} \left| \frac{\underline{\quad} \text{ mol } C_2H_5Cl \text{ produced}}{\text{mol } Cl_2 \text{ react}} = \underline{\quad} \text{ mol } C_2H_5Cl \right.$$

(4.50-11)

$$\Rightarrow \textbf{Fractional Yield} = \frac{n_{C_2H_5Cl}}{(n_{C_2H_5Cl})_{max}} = \underline{\qquad}$$

(d) Suppose the reactor is built and started up and the conversion is only 14%. Chromatographic analysis shows that there is no Cl_2 in the product but another species with a molecular weight higher than that of dichloroethane is present. Offer an explanation for these results. (*Hint:* If chloroethane can react with chlorine and form dichloroethane, what do you suppose might happen to dichloroethane if there is chlorine around? Write a stoichiometric equation to support your answer.)

(4.50-12)

PROBLEM 4.59

Ethylene oxide is produced by the catalytic oxidation of ethylene:

$$2C_2H_4 + O_2 \longrightarrow 2C_2H_4O$$

An undesired competing reaction is the combustion of ethylene:

$$C_2H_4 + 3O_2 \longrightarrow 2CO_2 + 2H_2O$$

The feed to the reactor (*not* the fresh feed to the process) contains 3 moles of ethylene per mole of oxygen. The single-pass conversion of ethylene is 20%, and for every 100 moles of ethylene consumed in the reactor, 90 moles of ethylene oxide emerge in the reactor products. A multiple-unit process is used to separate the products: ethylene and oxygen are recycled to the reactor, ethylene oxide is sold as a product, and carbon dioxide and water are discarded.

(a) Assume a quantity of the reactor feed stream as a basis of calculation, draw and label the flowchart, perform a degree-of-freedom analysis, and write the equations you would use to calculate (i) the molar flow rates of ethylene and oxygen in the fresh feed, (ii) the production rate of ethylene oxide, and (iii) the overall conversion of ethylene. Do no calculations.

Strategy

Let's choose a basis of 100 mol/h of feed to the reactor. From the given ethylene/oxygen ratio of 3:1, the feed must consist of 75 mol/h of ethylene and 25 mol/h of oxygen.

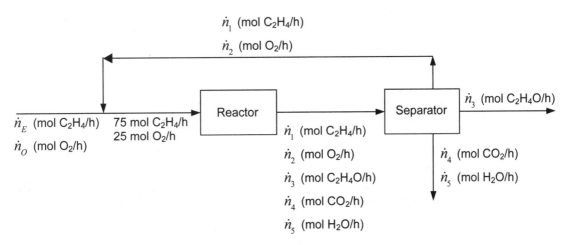

Since only 20% of the entering ethylene reacts, the product stream must contain unreacted ethylene and oxygen, product (ethylene oxide), and by-products (carbon dioxide and water). Although the separation section of the process has multiple units, we are given no information about them—all we can (and need) do is treat this part of the process as a single step which separates the product, the by-products, and the unconverted reactants. Since no reactions take place in the separator, we can justify labeling the streams leaving the separator with the same variables used for the stream that exits the reactor.

We will use degree-of-freedom analyses to plan the solution, using atomic species balances for reactive systems. (The alternative would be to use extents of reaction, which would work equally well.)

Name: _____

Date: _____

Solution

Complete the tables. (*Caution*: Be sure any given information applies to a sub-system before including the information in the degree-of-freedom analysis.)

(4.59-1)

DEGREE-OF-FREEDOM ANALYSIS: OVERALL PROCESS		
UNKNOWNS AND INFORMATION		JUSTIFICATION/CONCLUSION
+ ___ unknowns		
− ___ balances	_____	count atomic species balances
2 DOF		Don't begin with this system

(4.59-2)

Q: Why weren't the given single-pass conversion and ethylene oxide yield counted as additional relations in the overall process DOFA?

A: _____

(4.59-3)

DEGREE-OF-FREEDOM ANALYSIS: MIXING POINT		
UNKNOWNS AND INFORMATION		JUSTIFICATION/CONCLUSION
+ ___ unknowns		
− ___ balances	_____	Remember, this is a non-reactive system
2 DOF		Don't begin with this system

(4.59-4)

DEGREE-OF-FREEDOM ANALYSIS: SEPARATOR		
UNKNOWNS AND INFORMATION		JUSTIFICATION/CONCLUSION
+ 5 unknowns		
− 0 balances		
5 DOF		

(4.59-5)

Q: Why are there no allowable balances on the separator? (If you're unsure, try writing one.)

A: _____

(4.59-6)

DEGREE-OF-FREEDOM ANALYSIS: REACTOR		
UNKNOWNS AND INFORMATION		JUSTIFICATION/CONCLUSION
+ ___ unknowns		
− ___ balances	_____	count atomic species balances
− 1 _____		
− 1 _____		
0 DOF		Start here, determine listed variables

Since the reactor has zero degrees-of-freedom, we will begin with it and determine the values of $\dot{n}_1 - \dot{n}_5$. With $\dot{n}_1$ and $\dot{n}_2$ known, balances around the fresh feed-recycle mixing point will give us $\dot{n}_E$ and $\dot{n}_O$.

20% single-pass conversion of C_2H_4 $\Rightarrow 80\%$ of the C_2H_4 entering the reactor emerges from it unreacted. (Convert this statement into an equation.)

$$\widehat{\dot{n}_1}(\text{mol } C_2H_4/h) = \underline{\hspace{4cm}}$$ (4.59-7)

Yield: of the 20% of C_2H_4 that reacts, 90% forms C_2H_4O

$$\widehat{\dot{n}_3}\left(\frac{\text{mol } C_2H_4O \text{ produced}}{h}\right)$$ (4.59-8)

$$= \frac{75 \text{ mol } C_2H_4 \text{ fed}}{h} \left|\frac{\underline{\hspace{1cm}} \text{ mol } C_2H_4 \text{ react}}{\text{mol } C_2H_4 \text{ fed}}\right| \frac{\underline{\hspace{1cm}} \text{ mol } C_2H_4O \text{ produced}}{\underline{\hspace{1cm}} \text{ mol } C_2H_4 \text{ react}}$$

(4.59-9)

Reactor C balance: $\dfrac{75 \text{ mol } C_2H_4}{h} \left|\dfrac{2 \text{ mol C}}{1 \text{ mol } C_2H_4}\right. = \dot{n}_1(2) + \dot{n}_3(2) + \dot{n}_4$

Reactor H balance: _____

Reactor O balance: _____

Mixing point C_2H_4 balance: _____

Mixing point O_2 balance: _____

Overall C_2H_4 conversion:

$$X_E = \frac{(C_2H_4)_{\text{in fresh feed}} - (C_2H_4)_{\text{in final product}}}{(C_2H_4)_{\text{in fresh feed}}} = \underline{\hspace{4cm}}$$

(b) Calculate the quantities specified in Part (a), either manually or with an equation-solving program.

Let's use E-Z Solve. In the boxes below, write the set of equations for the unknown system variables and the overall conversion (4.59-7, 4.59-8, and the equations of 4.59-9) in E-Z Solve syntax, solve them using the program and record your results, and fill in your answers for the quantities requested in the problem statement.

Name: _____

Date: _____

(4.59-10)

//Problem 4.59, E-Z Solve Code

n1 = 0.80*75   //20%single-pass ethylene conversion

n3 = 75*0.20*0.9   //90% ethylene oxide yield

75*2 = n1*2 + n3*2 + n4   //Reactor C balance

75*4 = _____   //Reactor H balance

_____   //Reactor O balance

_____   //Mixing point C2H4 balance

_____   //Mixing point O2 balance

XE = _____   //Overall conversion

(4.59-11)

$\dot{n}_1 =$ _____ $\dot{n}_2 = 13.75$ mol O_2/h $\dot{n}_3 =$ _____ $\dot{n}_4 =$ _____

$\dot{n}_5 =$ _____ $\dot{n}_E = 15.0$ mol C_2H_4/h $\dot{n}_O =$ _____ $X_E=$ _____   (UNITS!)

(c) Calculate the molar flow rates of ethylene and oxygen in the fresh feed needed to produce 1 ton per hour of ethylene oxide.

Strategy

This is a scale-up calculation. The flow rate corresponding to the original basis of 100 mol fed to the reactor is $\dot{n}_3 = 13.5$ mol C_2H_4O/h. The newly specified mass flow rate of ethylene oxide can be converted to a molar flow rate, and a scale factor may be defined as the ratio of this flow rate to n_3. Any molar quantities calculated for the original basis may then be multiplied by the scale factor to determine the corresponding quantities for the new basis of 1 ton C_2H_5O/h.

Solution

$$(\dot{n}_3)_{new} = \frac{1 \text{ ton } C_2H_4O}{h} \left| \frac{2000 \text{ lb}_m}{1 \text{ ton}} \right| \frac{1 \text{ lb-mol } C_2H_4O}{44.05 \text{ lb}_m \ C_2H_4O} = 45.4 \frac{\text{lb-mol } C_2H_4O}{h}$$

Scale factor: $\dfrac{(\dot{n}_3)_{new}}{(\dot{n}_3)_{original}} = \dfrac{45.4 \text{ lb-mole/h}}{13.5 \text{ mol/h}} = 3.363 \dfrac{\text{lb-mole/h}}{\text{mol/h}}$

Ethylene feed rate: $(\dot{n}_E)_{new} =$ (4.59-12)

Oxgyen feed rate: $(\dot{n}_O)_{new} =$ (4.59-13)

PROBLEM 4.70

n-Pentane is burned with excess air in a continuous combustion chamber.

(a) A technician runs an analysis of the product gas and reports that it contains 0.270 mole% pentane, 5.3% oxygen, 9.1% carbon dioxide, and the balance nitrogen *on a dry basis*. Assume 100 mol of dry product gas as a basis of calculation, draw and label a flowchart, perform a degree-of-freedom analysis based on atomic species balances, and show that the system has −1 degrees of freedom. Interpret this result.

(b) Use balances to prove that the reported product gas composition could not possibly be correct.

(c) The technician reruns the analysis and reports new values of 0.304 mole% pentane, 5.9% oxygen, 10.2% carbon dioxide, and balance nitrogen. Verify that this result *could* be correct and, assuming that it is, calculate the percent excess air fed to the reactor and the fractional conversion of pentane.

Strategy

This problem illustrates an important fact of life in any industrial setting: engineers should not accept reported measured values of process variables on faith. Always check the values by (among other things) determining whether the results are confirmed by material balances on the process. You will often hear this exercise referred to as "closing the mass balance." If the balance doesn't close, the data are either in error or they have been reported incorrectly.

Solution

(a) The flowchart is shown below. Can you tell by looking at it what the ratio of atomic oxygen (O) to atomic nitrogen (N) in the product stream must be?

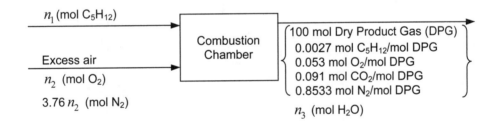

Your turn. Complete the DOFA.

(4.70-1)

DEGREE-OF-FREEDOM ANALYSIS: OVERALL PROCESS	
UNKNOWNS AND INFORMATION	JUSTIFICATION/CONCLUSION
+___ unknowns	
−___ reactive atomic species balances	
− 1 inert species balance	
−1 DOF	The system is *overspecified*; there are more equations than unknowns.

We have a problem; there are three unknown variables and four equations relating them. There are two possibilities:

- One additional process variable was forgotten when the flowchart was drawn. That is not the case in this process—all process species have been identified in the inlet and outlet streams and their amounts can be expressed in terms of the numbers and variables on the chart. (Convince yourself.)

- One more process variable value was specified than was necessary to determine all remaining unknowns, which is what has happened. There are also two possibilities in this situation: either the specified variable values satisfy all of the balance equations or they don't. To see which it is, determine all three unknowns by solving three of the four possible equations, and then substitute the results into the fourth equation to see if that equation is satisfied. If it is, all of the system balances close and the specified and calculated variable values can be accepted; otherwise, one or more of the specified values must be incorrect.

(b) Use balances to prove that the reported product gas composition could not possibly be correct.

Solution

Your turn again. Write the balance equations and circle the unknown to be solved for. When you get to the oxygen balance, you should find that the variables involved (n_2 and n_3) have already been determined. Use E-Z Solve or your calculator to determine the values of the variables after you finish the hydrogen balance and record the values. Substitute the calculated variable values into the oxygen balance and see if it closes.

(4.70-2)

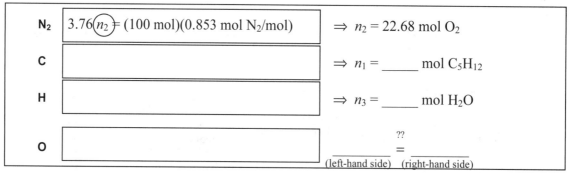

N₂	$3.76\,\widehat{n_2} = (100\ \text{mol})(0.853\ \text{mol N}_2/\text{mol})$	$\Rightarrow n_2 = 22.68\ \text{mol O}_2$
C		$\Rightarrow n_1 = \underline{\quad}\ \text{mol C}_5\text{H}_{12}$
H		$\Rightarrow n_3 = \underline{\quad}\ \text{mol H}_2\text{O}$
O		$\dfrac{??}{\underline{\text{(left-hand side)}}} = \underline{\text{(right-hand side)}}$

Did the mass balance close?   yes ☐   no ☐

(4.70-3)

(c) The technician reruns the analysis and reports new values of 0.304 mole% pentane, 5.9% oxygen, 10.2% carbon dioxide, and balance nitrogen. Verify that this result *could* be correct and, assuming that it is, calculate the percent excess air fed to the reactor and the fractional conversion of pentane. The corrected flowchart follows.

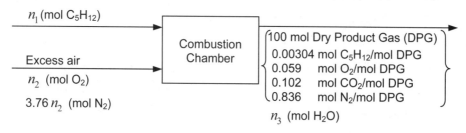

(4.70-4)

N$_2$		$\Rightarrow n_2 =$ _____
C		$\Rightarrow n_1 =$ _____
H		$\Rightarrow n_3 =$ _____
O		$\dfrac{??}{\text{_____}} = $ _____

Did the mass balance close this time?   yes ☐   no ☐   **(4.70-5)**

Using the values just calculated, determine the fractional conversion of C_5H_{12}, the theoretical oxygen required for the process, and the percent excess air. Show your calculations and box your answers.

(4.70-6)

Fractional C$_5$H$_{12}$ Conversion:

Theoretical O$_2$:

% Excess Air:

Notes and Calculations

Thus far, you have been given either mass or molar quantities (and mass or mole fractions) to describe stream flows in material balance problems. In Chapter 5, you encounter more realistic problem descriptions in which stream flows are specified in terms of volumetric quantities or flow rates, and your task is then to determine the density of the stream material and use this value to calculate the mass or molar quantity or flow rate of the stream. The mass or molar quantities, in turn, are used in material balance calculations. When the materials involved are solids and liquids, the densities can normally be assumed independent of temperature and pressure and looked up in published tables (e.g., Table B.1 in your text). When the streams are gases, however, density depends strongly on temperature and pressure, and an equation of state (EOS) must be used to relate molar and volumetric quantities. The simplest EOS is the ideal gas equation of state, which works well at high temperatures and low pressures. At other conditions, a more complex EOS must be used.

Be sure to read Chapter 5 in your text and work through the example problems before tackling the problems in the workbook. The text explains the concepts; we will work problems without as much explanation.

PROBLEM 5.10

A stream of air enters a 7.50 cm pipe at a velocity of 60.0 m/s at 27°C and 1.80 bar (gauge). At a point downstream, the air flows through a 5.00 cm ID pipe at 60°C and 1.53 bar (gauge). What is the velocity of the gas at this point?

Strategy

The problem statement doesn't ask for it, but it is always advisable to draw and label a flowchart first. The air is flowing at steady-state so the mass flow rate in each pipe must be the same and, since there is no chemical reaction, the molar flow rate must be the same as well.

$\dot{n}_o$ (kg air/s)
7.50 cm ID
5.00 cm ID
$\dot{n}_o$ (kg air/s)

60.0 m/s
27°C
1.80 bar (gauge)
$\dot{V}_1$ (m³/s)

u (m/s)
60°C
1.53 bar (gauge)
$\dot{V}_2$ (m³/s)

We have built the mole balance into the flowchart by using $\dot{n}_o$ for both inlet and outlet flow rates.

To solve the problem, we need to recognize the need for two relationships. The first relates the velocity of the gas (which we know at the inlet) to its volumetric flow rate at that point, $\dot{V} = uA_x$ where A_x is the cross-sectional area of the pipe. Also, we need an equation to relate the volumetric flow, temperature, and pressure of the gas to its molar flow rate. We can use the ideal gas equation of state (EOS) for the second relationship, but we should test its validity for the stated conditions of the gas.

We will carry out the calculation in several steps:

(1) Determine whether the air can be assumed to behave as an ideal gas at the conditions in the 7.50 cm ID and the 5.00 cm ID pipes.
(2) Calculate the volumetric flow rate of the gas in the 7.50 cm ID pipe.
(3) If the air behaves as an ideal gas in the 7.50 cm ID pipe, calculate the molar flow rate, $\dot{n}_o$.
(4) If the air behaves as an ideal gas in the 5.00 cm ID pipe, calculate the volumetric flow rate of the air in the 5.00 cm ID pipe from the molar flow rate.
(5) Calculate the velocity of the air in the 5.00 cm ID pipe from the volumetric flow rate and the cross-sectional area.

Solution

Let's do step (1) and make sure we are justified in treating the air as an ideal gas. Review the criteria on p. 192 of your text and determine the molar volume of the air at each condition. Can the air be treated as an ideal gas in both pipes?   (5.10-1)

7.50 cm ID	5.00 cm ID

$$\hat{V}_{ideal} = \frac{RT}{P} = \frac{\underline{\qquad} \; \dfrac{L \cdot bar}{mol \cdot K} \; \left| \; \underline{\qquad} \; K\right.}{\underline{\qquad} \; bar}$$

$$= \underline{\qquad} \frac{L}{mol} (\leq, >) \; 5 \frac{L}{mol}$$

Treat air as ideal?   ❐ Yes   ❐ No

Treat air as ideal?   ❐ Yes   ❐ No

Now, complete the solution by doing the calculations for steps (2)–(5).   (5.10-2)

(2) $\dot{V}_{7.5\,cm} \left(\dfrac{m^3}{s} \right) =$

(3) $\dot{n}_0 \left(\dfrac{mol}{s} \right) = \dfrac{P\dot{V}}{RT} =$

(4) $\dot{V}_{5.0\,cm} \left(\dfrac{m^3}{s} \right) =$

(5) $u_{5.0\,cm} \left(\dfrac{m}{s} \right) =$

$$= \underline{165.7} \; m/s$$

Instead of doing all that, you could calculate the outlet velocity with a single dimensional equation that is easy to derive once you completely understand the problem solution. Here's the equation—derive it in the box below and prove that you get the same result for the velocity in the 5.00 cm ID pipe as you did above.

$$\frac{u_2}{u_1} = \frac{T_2}{T_1} \cdot \frac{P_1}{P_2} \cdot \frac{D_1^2}{D_2^2}$$

where subscripts 1 and 2 refer to conditions in the 7.50 cm ID and 5.00 cm ID pipes, respectively.

(5.10-3)

Obviously, this problem is straightforward once you recognize how many and which relationships you need to solve it. Your recognition skills will get much better as you work more problems. Until then, you can help yourself by doing a degree-of-freedom analysis even if the problem statement doesn't call for one. The DOFA always *anticipates* the number of relationships needed to solve a problem—knowing how many relationships are needed is half the battle. For example, the DOFA for this problem is:

DEGREE-OF-FREEDOM ANALYSIS		
UNKNOWNS AND INFORMATION		JUSTIFICATION/CONCLUSION
+ 4 unknowns	$\dot{n}_o,\, u,\, \dot{V}_1,\, \dot{V}_2$	
−2 PVT relationships	one for each pipe	
−2 velocity-area relationships	one for each pipe	
0 DOF		Problem is solvable

Note: The Ideal Gas Equation of State and When NOT to Use It

Before we leave this problem, a bit of discussion about the ideal gas EOS is in order. The criteria on p. 192 of your text are used to determine whether or not the ideal gas EOS might apply to a given gas at a given temperature and pressure. If the appropriate molar density criterion fails, the ideal gas EOS cannot be used with confidence. If the molar density criterion holds, the ideal gas EOS *might* be applicable. Even so, for certain molecules, the ideal gas EOS should generally not be used for more than rough estimations since these molecules exhibit strong interactions even at the high molar volumes used as benchmarks in the criteria.

As an example, consider water vapor at 300°C and 1 bar. If you calculate RT/P, the value is 54.48 L/mol. This is sufficiently higher than the 20 L/mol criterion that you would assume water vapor at this high temperature and low pressure would act like an ideal gas. And, you would be wrong. The actual molar volume of water vapor at 300°C and 1 bar is 47.56 L/mol[5.1]—14.5% lower than expected from the RT/P calculation!

Water vapor is almost always non-ideal because of very strong hydrogen bonding. You can expect similar behavior, albeit not quite as pronounced, for vapors of organic acids and alcohols. Fortunately, we have tables available for water with very accurate physical property data (such as Tables B.5–B.7 of your text). The same cannot be said for most organic acids and higher carbon-number alcohols, for which one of the non-ideal equations of state that come later in Chapter 5 are better used.

The assumption of ideal gas behavior is probably the most abused of assumptions typically used in engineering calculations. When solving problems in this book, as long as the criteria on p. 192 are satisfied you may go ahead and use the ideal gas EOS unless you are specifically directed to do otherwise, but if you are solving PVT problems in the workplace and need a reasonable amount of accuracy in your results, don't fall into the ideal gas trap.

[5.1] Find $\hat{V} = 2.64$ m³/kg at 300°C and 1 bar in Table B.7 of the text and convert the units. The molecular weight of water (Table B.1) is 18.016 g/mol.

PROBLEM 5.19

Sax and Lewis[5.2] describe the hazards of breathing air containing appreciable amounts of an asphyxiant (a gas that has no specific toxicity but, when inhaled, excludes oxygen from the lungs). When the mole percent of the asphyxiant in the air reaches 50%, marked symptoms of distress appear, and at 75%, death occurs in a matter of minutes.

A small storage room whose dimensions are 2 m × 1.5 m × 3 m contains a number of expensive and dangerous chemicals. To prevent unauthorized entry, the room door is always locked and can be opened with a key from either side. A cylinder of liquid carbon dioxide is stored in the room. The valve on the cylinder is faulty and some of the contents have escaped over the weekend. The room temperature is 25°C.

(a) If the concentration of CO_2 reached the lethal 75 mole% level, what would be the mole percent of O_2?

Strategy

Mass balance calculations are sometimes simplified by combining two or more components that always remain together in the same proportion and treating the combination as if it were a single species. This technique is often used when air is involved in a calculation. The solution to part (a) is simplified by first treating air as one of two components in the room (the other is the CO_2) and then using the composition of air to determine the amount of O_2.

Solution

When the mole fraction of CO_2 equals 0.75, the mole fraction of air must equal 0.25. Air may be considered to contain 21 mole% O_2 and 79 mole% N_2.[5.3] Take a basis of 1 mol of gas with $y_{CO_2} = 0.75$ and compute the mole percent of O_2.

(5.19-1)

$$\text{mole\% O}_2 = \frac{1 \text{ mol gas}}{} \left| \frac{\text{mol air}}{\text{mol gas}} \right| \frac{\text{mol O}_2}{\text{mol air}} \times 100\% = \underline{\qquad}$$

(b) How much CO_2 (kg) is present in the room when the lethal concentration is reached? Why would more than that amount have to escape from the cylinder for this concentration to be reached?

Strategy

We know the pressure (1 atm), the temperature (25°C), and the dimensions of the room—which gives us the volume. With P, T, and V, we can compute n using the ideal gas EOS. To find the total moles of CO_2 in the room, we need to multiply the total number of kmol of gas in the room by the fraction that is CO_2. The molecular weight of CO_2 is then used to convert from kmol to kg. Before doing the calculation, check to make sure the ideal gas EOS is a reasonable approximation by computing RT/P.

Solution

$$\frac{RT}{P} =$$

(5.19-2)

[5.2] N. I. Sax and R. J. Lewis, *Hazardous Chemical Desk Reference*, Van Nostrand Reinhold, New York, 1987.
[5.3] Yes, there are trace amounts of other gases but we ignore their presence in most engineering calculations.

According to the criterion on p. 192 of the text, is the ideal gas assumption reasonable to apply in this problem? ❐ Yes   ❐ No   (5.19-3)

Use a single dimensional equation to estimate the mass of CO_2 in the room when the mole fraction is 0.75.

$$= \underline{12.1 \text{ kg } CO_2} \qquad \text{(5.19-4)}$$

    For the second question in part (b), (Why would more than the calculated amount have to escape from the cylinder for this concentration to be reached?), here's a hint. When several people spend a very long time in a typical room, they do not suffocate even though they are consuming oxygen and exhaling carbon dioxide. Why not? Write a complete sentence to answer the question in the box below.

                                                                          (5.19-5)

(c) Describe a set of events that could result in a fatality in the given situation. Suggest at least two measures that would reduce the hazards associated with storage of this seemingly harmless substance.

Solution

    Come up with a *realistic* scenario and two procedures you would put in place to prevent a terrible accident from happening.

                                                                          (5.19-6)

PROBLEM 5.23

A balloon 20.0 m in diameter is filled with helium at a gauge pressure of 2.0 atm. A man is standing in a basket suspended from the bottom of the balloon. A restraining cable attached to the bucket keeps the balloon from rising. The balloon (not including the gas it contains), the basket, and the man have a combined mass of 150 kg. The temperature is 24°C that day, and the barometer reads 760 mm Hg.

(a) Calculate the mass (kg) and weight (N) of the helium in the balloon.

Strategy

Often, it helps to start with the result you want and work your way backward through the problem. Let's try that technique here. To determine the weight (force) of helium, we need the mass of helium. Pure helium is in the balloon so the mass of helium is easily found if we determine first how many moles of helium are in the balloon. To find the moles of helium, we need the pressure, temperature, and volume of the gas. We know the pressure and temperature and, assuming the balloon is spherical, we can determine the volume geometrically.

Solution

The procedure (in the forward direction) is below. You do the calculations.

(1) Calculate the volume, V_b, of a spherical balloon 20.0 m in diameter.
(2) Calculate the moles of helium, n_{He}, from the volume, temperature, and pressure.
(3) Calculate the mass of helium, m_{He}, from the moles of helium.
(4) Calculate the weight of helium, w_{He}, from the mass of helium (see Section 2.4).

(5.23-1)

Volume of balloon: V_b =	= _____ m^3
Moles He in balloon: n_{He} =	= _____ kmol He
Mass of He in balloon: m_{He} =	= 2065 kg He
Weight of He in balloon: W_{He} =	= _____ N

(b) How much force is exerted on the balloon by the restraining cable? (*Recall*: The buoyant force on a submerged object equals the weight of the fluid—in this case, the air—displaced by the object. Neglect the volume of the basket and its contents.)

Strategy

Whenever an object is motionless, there can be no net force acting on it. (If there were one, what would happen according to Newton?) The key to problems like this one is to draw a free body diagram, showing all the forces acting on the object, and then use a force balance to find the unknown force (in this case, the force exerted by the restraining cable).

Name: _____

Date: _____

Solution

The sketch below shows the vectors for the bouyant force, the weight of the balloon and its attachments (including the helium), and the force imposed by the restraining cable.

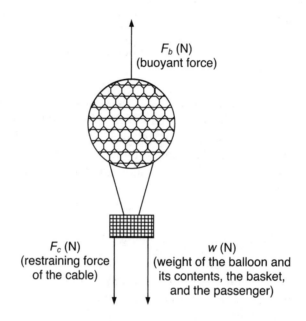

F_b (N)
(buoyant force)

F_c (N)
(restraining force
of the cable)

w (N)
(weight of the balloon and
its contents, the basket,
and the passenger)

The combined mass of the balloon (*not* including the helium), basket, and passenger is 150 kg. From Eq. (2.4-4) in the text, the corresponding weight and the total weight of the balloon including the gas is

$$w_b = m_b\,(g) = \frac{150\,\text{kg}}{} \left| \frac{\underline{}\,\text{m}}{s^2} \right| \frac{\text{N}}{\dfrac{\text{kg}\cdot\text{m}}{s^2}} = \underline{}\,\text{N}$$

(5.23-2)

$$\Rightarrow w = w_{gas} + w_b = \underline{}\ \text{N} + \underline{}\ \text{N} = 21,720\ \text{N}$$

Use the volume of the balloon (neglecting the volume of the basket and its contents) to find the moles, mass, and weight of the air displaced, and solve the force balance for the restraining force of the cable. Take the molecular weight of air to be 29.0 kg/kmol .

(5.23-3)

Moles of air displaced (kmol) =

Mass of air displaced (kg) =

(5.23-4)

Weight of air displaced (N): $F_b =$
(= the buoyant force)

Cable restraining force (N): $F_c =$

(c) Calculate the initial acceleration of the balloon when the restraining cable is released.

Strategy

When the restraining cable is released, a net upward force on the balloon results. Calculate it and find the acceleration from Newton's second law of motion.

Solution
(5.23-5)

Net upward force (N) =

Acceleration (m/s^2) =

(d) Why does the balloon eventually stop rising? What would you need to know to calculate the altitude at which it stops?

(5.23-6)

(e) Suppose at its point of suspension in midair the balloon is heated, raising the temperature of the helium. What happens and why? Answer below.

(5.23-7)

Notes and Calculations

PROBLEM 5.38
Propylene is hydrogenated in a batch reactor:

$$C_3H_6(g) + H_2(g) \longrightarrow C_3H_8(g)$$

Equimolar amounts of propylene and hydrogen are fed into the reactor at 25°C and a total absolute pressure of 32.0 atm, and some time later the temperature is 235°C. You may assume ideal gas behavior for this problem, although at the high pressures involved this assumption constitutes a crude approximation at best.

(a) If the reaction goes to completion at 235°C, what would be the final pressure?
(b) If the pressure is 35.1 atm and the temperature is 235°C, what percentage of the propylene fed has reacted?
(c) Construct a graph of pressure versus fractional conversion of propylene, assuming $T = 235°C$. Use a graph to confirm the results in parts (a) and (b). (Note: We'll use E-Z Solve for this part.)

Strategy
Since we are not told how much of the reaction mixture is charged, we'll choose a basis of 100 mol C_3H_6, and since the feed is equimolar in propylene and hydrogen there must also be 100 mol H_2. We know the extent of reaction but not the pressure in part (a) and vice versa in part (b), so we label the chart as though neither is known. Also, we assume that the reaction volume is the same throughout the reaction. (Put another way, we assume the reactor is a rigid vessel and not an expandable balloon or a cylinder with a movable piston.)

(a) If the reaction goes to completion at 235°C, what would be the final pressure?

Solution
If the reaction is complete, we can easily determine the three unknown molar quantities in the outlet stream labeling. (*Hint:* Use the expression for n_1 to find the extent and then use the extent to find n_2 and n_3).

$$n_1 = 0 \text{ mol } C_3H_6, \quad n_2 = \underline{\quad\quad} \text{ mol } H_2, \quad n_3 = \underline{\quad\quad} \text{ mol } C_3H_8 \qquad \textbf{(5.38-1)}$$

We are left with two unknowns (V and P_f), and we can write two equations to determine them (the ideal gas equation-of-state at the inlet and outlet). Since we are not interested in V, the most efficient solution method is to eliminate it between the two equations and solve for the final pressure.

Initial: $(32 \text{ atm})V = (200 \text{ mol})R(298K)$ $\quad\xrightarrow[\text{by 1}^{\text{st}}]{\text{Divide 2}^{\text{nd}}}\quad \dfrac{P_f}{32 \text{ atm}} = \dfrac{\rule{1cm}{0.4pt}}{\rule{1cm}{0.4pt}} \Rightarrow P_f = \underline{\quad} \text{ atm} \qquad \textbf{(5.38-2)}$
Final: $P_f V = (100 \text{ mol})R(508K)$

Name: _____

Date: _____

(b) If the pressure is 35.1 atm and the temperature is 235°C, what percentage of the propylene fed has reacted?

Solution

Add the pressure to the flowchart and complete the DOFA.

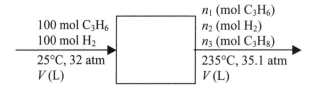

n_1 (mol C_3H_6)
n_2 (mol H_2)
n_3 (mol C_3H_8)

235°C, 35.1 atm
V (L)

100 mol C_3H_6
100 mol H_2

25°C, 32 atm
V (L)

(5.38-3)

DEGREE-OF-FREEDOM ANALYSIS		
UNKNOWNS AND INFORMATION		CONCLUSION
+ 4 unknowns		
–___ atomic species balances		
–___ PVT relationships		
0 DOF		Problem is solvable

Writing the gas law equations and atomic species balances will enable us to solve for n_1, n_2, n_3, and V, and we may then calculate the percentage conversion of C_3H_8 from the relationship

$$p_c = \frac{(100 - n_1) \text{ mol } C_3H_8 \text{ reacted}}{100 \text{ mol } C_3H_8 \text{ fed}} \times 100\%$$

Write the five equations needed (but don't do any of the algebra yet).   **(5.38-4)**

C balance: $(100)(3) = 3n_1 + 3n_3$

H balance:

Gas law at inlet: $(32 \text{ atm})V(L) = (200 \text{ mol})(0.08206 \frac{\text{L·atm}}{\text{mol·K}})(298\text{K})$

Gas law at outlet:

Percentage conversion:

It would not be too difficult to solve these equations manually, but it is even easier to use E-Z Solve. Write the program below (the first and last lines are given), enter and run it. The final fractional conversion you should obtain is given below so you can verify your answer.

(5.38-5)

```
//E-Z Solve program for Problem 5.38(b)
100*3 = 3*n1 + 3*n3
_____
_____
_____
pc = 100 – n1          // simplified form of ((100 – n1)/100)*100
```

Solution: $p_c = \underline{71}\%$

(c) Construct a graph of pressure versus fractional conversion of propylene, assuming $T = 235°C$. Use the graph to confirm the results in parts (a) and (b).

Strategy

There are a number of ways you can approach this problem, including these (in increasing order of convenience).

(1) Substitute a number of different values of P_f into the equations you wrote for (**5.38-4**), solve for p_c manually, and draw the graph.

(2) Eliminate all variables but p_c and P_f from the equations in (**5.38-4**), derive an expression for P_f as a function of p_c, and use a spreadsheet to generate the points (e.g., determine p_f for $0 \leq p_c \leq 100$ at intervals $\Delta p_c = 5$) and plot the graph.

(3) Do it all with E-Z Solve. You can have the program do a *parameter sweep* in which it varies the value of p_c from 0 to 100 in fixed steps of any interval you choose, calculates P_f for each value, and then plots the desired graph. This is the method we'll take you through. All of the steps listed below are found in the excellent tutorial accessible under the E-Z Solve Help menu.

Solution

- Replace the fixed value of 35.1 (for P_f) in the set of equations entered previously into E-Z Solve with the variable name Pf and add a sixth line defining an initial value of p_c (pc = 0 will work).
- Click the calculator icon or select "Solve/Sweep" under the Solutions menu. Click on the "Sweep" tab near the top of the window (not the one near the bottom). In the new window that comes up, make sure "Single sweep" and "Add new run" are checked.

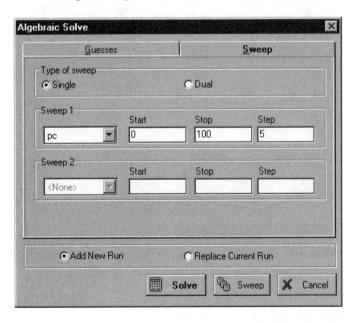

- Under "Sweep 1," enter "pc" (if that's what you called the percentage conversion variable), start at 0, stop at 100, and use a step size of 5. Once these values are entered, click the "Sweep" tab near the bottom of the window. A table of solutions for each specified value of p_c should appear. Save.

- We could transfer the values of P_f and p_c to a spreadsheet and do the rest that way, but let's not stop now. Under the Solutions menu, select 2D Graph → New. The Edit 2D Graph dialog appears. Select p_c as the X-axis variable and P_f as the Y-axis variable. Put a check mark in the "Show grid" box, and click on the "Attributes" tab.

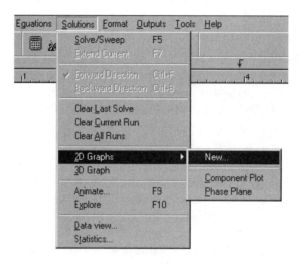

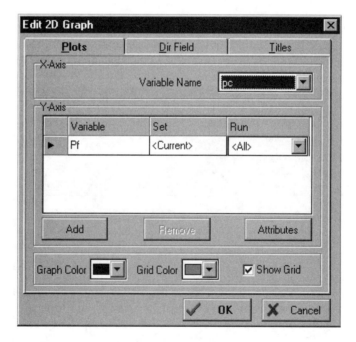

- In the Plot Attributes dialog, make sure "Show" is checked, change the line color if you wish to (the background of the graph will be black), and leave the rest alone for now. Click "OK."

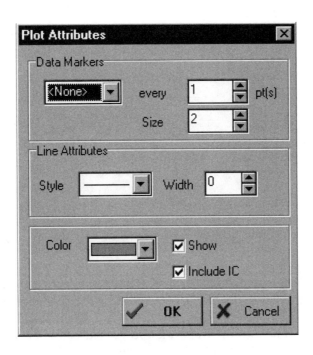

- Back in the Edit 2D Graph dialog, click on the "Titles" tab at the top. For "X-axis Title" enter "Percentage Conversion of Propylene" and for "Y-axis Title" enter "Reactor Pressure (atm)". Click OK, and OK again when the Edit window returns. The desired plot should appear. Save.

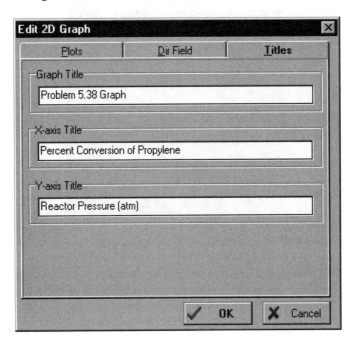

- The Y-axis scale may make it difficult to read values of the pressure, so we'll change it. Right-click on the graph, select "Scales" and the Graph Scales dialog should appear. Change the X-axis tick label interval from 2 to 1. (The picture below shows this view of the dialog.) Click on the "Y-Scale" tab at the top. Deselect "Auto Scale." Have the scale run from 0 to 100 and change the tick label interval from 2 to 1. Click OK. Save.

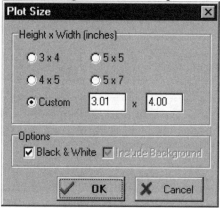

- Finally, have E-Z Solve prepare a report. Select the complete program text in the workspace, copy it, click on the "Report" tab at the bottom of the E-Z Solve window, and paste the program into it. Add any text you wish to add (such as your name).
- Click on the "A" tab next to the Report tab to get back to your program and edit the graph again (right-click on it and select "Edit"). Change the graph color to white, the grid color to dark gray, and the line color of the plot (under "Attributes") to black, and click OK. (If you didn't do this, all you'd see on the report page is a black square.) Right click on the graph again and select "Copy." Click "Custom" and enter 2.23" x 4.00" as the size, black and white as the option, and click OK[5.4]. Then go back to the report file and paste the graph into it. Save.

- Click on the "A" tab. Select the entire data table by clicking on the top left cell, then, holding the shift key down, on the bottom right cell. Copy the cells. Go back to Report, and paste the data table below the graph. If some of the 0 values are reported as something like −3.55271E−15, change them to 0 to improve the appearance of the table.
- Either (a) print the report file, or (b) export it (under the "File" menu) to a rich text format file that you can edit with a word processing program. The final result should appear as below.

[5.4] There is a weird bug in E-Z Solve that affects whether the x-axis label shows up properly when pasting a graph into an E-Z Solve report. No matter what width is chosen for the copied graph, you must vary the height to find one that puts the label in the right location. If the height is too large, the label will move up into the graph, and if it is too small, the label will be cut off or disappear. Values that have worked for us are 2.23 and 3.01—but it seems to vary on different computers so you will have to experiment on yours. After many years of use, this is the only bug we've ever found in the program.

```
// E-Z Solve program for Problem 5.38(b)
300 = 3*n1 + 3*n3
800 = 6*n1 + 2*n2 + 8*n3
32.0*V = 200*0.08206*298
Pf*V = (n1+n2+n3)*0.08206*508
f = 100-n1
f = 0
```

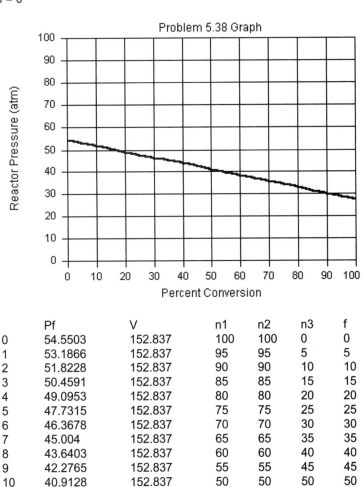

Problem 5.38 Graph

	Pf	V	n1	n2	n3	f
0	54.5503	152.837	100	100	0	0
1	53.1866	152.837	95	95	5	5
2	51.8228	152.837	90	90	10	10
3	50.4591	152.837	85	85	15	15
4	49.0953	152.837	80	80	20	20
5	47.7315	152.837	75	75	25	25
6	46.3678	152.837	70	70	30	30
7	45.004	152.837	65	65	35	35
8	43.6403	152.837	60	60	40	40
9	42.2765	152.837	55	55	45	45
10	40.9128	152.837	50	50	50	50
11	39.549	152.837	45	45	55	55
12	38.1852	152.837	40	40	60	60
13	36.8215	152.837	35	35	65	65
14	35.4577	152.837	30	30	70	70
15	34.094	152.837	25	25	75	75
16	32.7302	152.837	20	20	80	80
17	31.3664	152.837	15	15	85	85
18	30.0027	152.837	10	10	90	90
19	28.6389	152.837	5	5	95	95
20	27.2752	152.837	0	0	100	100

Notes and Calculations

Name: _____

Date: _____

PROBLEM 5.46

A stream of liquid n-pentane flows at a rate of 50.4 L/min into a heating chamber, where it evaporates into a stream of air 15% in excess of the amount needed to burn the pentane completely. The temperature and gauge pressure of the entering air are 336K and 208.6 kPa. The heated gas flows into a combustion furnace in which a fraction of the pentane is burned. The product gas, which contains all of the unreacted pentane and no CO, goes to a condenser in which both the water formed in the furnace and the unreacted pentane are liquefied. The uncondensed gas leaves the condenser at 275K and 1 atm absolute. The liquid condensate is separated into its components, and the flow rate of the pentane is measured and found to be 3.175 kg/min.

(a) Calculate the fractional conversion of pentane achieved in the furnace and the volumetric flow rates (L/min) of the feed air, the gas leaving the condenser, and the liquid condensate before its components are separated.

Strategy

As we have said several times, the key to solving most of the problems in the text is to draw a flowchart, label it completely, and use it to keep track of what you know and what remains to be determined. At this point in the text, the processes described in the problems are becoming more and more complex. If you're finding it difficult to convert the verbal descriptions into flowcharts, you are not alone. Most students struggle with this task until they have done it often enough to become proficient. Many students find that the following four-step method helps them achieve that proficiency.

(1) The first time you read through the description, look only for the different process elements involved—reactors, absorbers, extractors, condensers, heaters, crystallizers etc. As you encounter these, draw and label them on your paper.

(2) Read the problem again. This time, consider the material streams involved in the process—feed streams from outside the process going to process units, product streams going from units to the outside, and streams going from one unit to another. Each time material "moves" in the process description, draw the appropriate arrow on your flowchart.

(3) Read the description one last time. Your goal with this reading is to write all information given about the process materials (flow rates, mole or mass fractions, phases, temperatures, and pressures) on the corresponding streams on the chart.

(4) The final tasks are to choose a basis of calculation if one is not specified in the process statement and to make sure that each stream is fully labeled, defining variables as needed so that the flow rate of every stream component can be expressed in terms of labeled variables and numbers.

Let's build the flowchart for this problem using this systematic approach. We strongly recommend that you make the effort to fill in the blank spaces as directed rather than jumping ahead to see the results. Once you've tried something yourself, whether you get it right or not, you'll understand the worked-out solution in a way that you never can if you simply look at the solution.

Name: _____

Date: _____

Creating a Flowchart from a Process Description

Here's the process description again.

"A stream of liquid n-pentane flows at a rate of 50.4 L/min into a heating chamber, where it evaporates into a stream of air 15% in excess of the amount needed to burn the pentane completely. The temperature and gauge pressure of the entering air are 336K and 208.6 kPa. The heated gas flows into a combustion furnace in which a fraction of the pentane is burned. The product gas, which contains all of the unreacted pentane and no CO, goes to a condenser in which both the water formed in the furnace and the unreacted pentane are liquefied. The uncondensed gas leaves the condenser at 275K and 1 atm absolute. The liquid condensate is separated into its components, and the flow rate of the pentane is measured and found to be 3.175 kg/min."

Step (1): Identify the process elements.

Draw and label boxes to represent the process units identified in the description.

Step (2): Draw the flow streams going to, from, and between the process elements.

You should have four boxes on your flowchart. (Remember that if the liquid effluent from the condenser is separated into its components, the separation has to take place in something.) Fix your chart if necessary, reread the process description, and draw the appropriate streams on your flowchart.

Step (3): Add information given for each stream(s).

Go through the process description once more and put all given information about process streams on your flowchart. Then compare your chart to the one shown on the next page. Don't worry if your layout is different—just make sure all of the process units and streams have the right relationships to one another on your flowchart and all of the relevant information is correctly labeled.

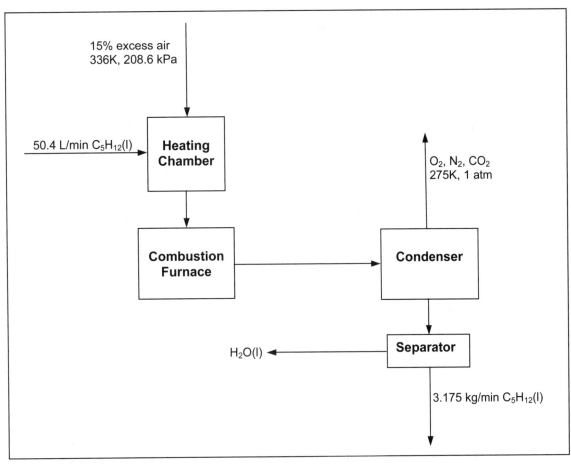

Step (4): Label unknowns for each stream.

Finally, label the above flowchart completely, defining unknowns so that the molar flow rates of all components of each stream can be expressed in terms of the chart labeling. For now, don't bother labeling the streams between the heating chamber and the combustion furnace and between the condenser and the separator. Here's the problem statement again for ease of reference.

"A stream of liquid n-pentane flows at a rate of 50.4 L/min into a heating chamber, where it evaporates into a stream of air 15% in excess of the amount needed to burn the pentane completely. The temperature and gauge pressure of the entering air are 336K and 208.6 kPa. The heated gas flows into a combustion furnace in which a fraction of the pentane is burned. The product gas, which contains all of the unreacted pentane and no CO, goes to a condenser in which both the water formed in the furnace and the unreacted pentane are liquefied. The uncondensed gas leaves the condenser at 275K and 1 atm absolute. The liquid condensate is separated into its components, and the flow rate of the pentane is measured and found to be 3.175 kg/min."

Verify that each labeled stream meets the criterion for being fully labeled, remembering that we will be writing mole balances and so the component molar flow rates must be expressible in terms of the labeling. The problem statement also requires us to calculate (and therefore label) the volumetric flow rates of the inlet air, the condenser outlet gas, and the combined liquid condensate. When you are satisfied, turn to the next page and compare your flowchart to ours.

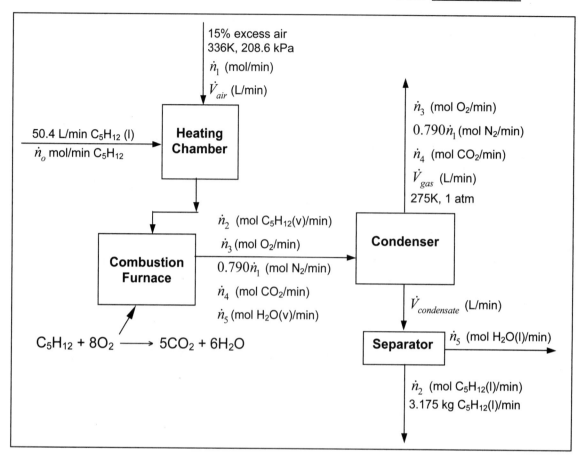

Before continuing, go back and see what (if anything) you missed in labeling your chart.

(a) Calculate the fractional conversion of pentane achieved in the furnace and the volumetric flow rates (L/min) of the feed air, the gas leaving the condenser, and the liquid condensate before its components are separated.

Solution

Let's begin by analyzing the overall process. Complete the DOF table.

(5.46-1)

DEGREE-OF-FREEDOM ANALYSIS OF OVERALL PROCESS	
UNKNOWNS AND INFORMATION	**JUSTIFICATION/CONCLUSION**
+ 8 unknowns _____	
− ____ atomic species balances _____	
− 1 _____	Calculate $\dot{n}_0$
− 1 _____	Calculate $\dot{n}_1$
− ____ ideal gas EOS	Calculate _____
− 1 _____	Calculate $\dot{n}_2$
0 DOF	Solve for all unknowns

(5.46-2)

Q: Why didn't we count a nitrogen balance as one of the equations relating the unknown variables?

A: _____

_____

The only quantity requested in the problem statement that is not calculated by doing an overall analysis is the volumetric flow rate of the liquid leaving the condenser, $\dot{V}_{condensate}$. That calculation is straightforward, however, once the flow rate of water in that stream ($\dot{n}_5$) has been determined (the mass flow rate of pentane in the stream is known).

Next, write the overall system equations in an efficient order, circling the variables for which you would solve (we've left some for you to circle). Do not do the algebra.

(5.46-3)

Pentane feed rate $\textcircled{$\dot{n}_0$}(\dfrac{\text{kmol } C_5H_{12}}{\text{min}}) = \dfrac{50.4 \text{ L}}{\text{min}} \left| \dfrac{\underline{\quad} \text{ kg}}{\text{L}} \right| \dfrac{1 \text{ kmol}}{\underline{\quad} \text{ kg}}$

Air feed rate $\dot{n}_1 (\dfrac{\text{kmol air}}{\text{min}}) = \dfrac{\dot{n}_0(\text{kmol } C_5H_{12})}{\text{min}} \left| \dfrac{\underline{\quad} \text{ kmol } O_2}{\text{kmol } C_5H_{12}} \right| \dfrac{\text{kmol air}}{\underline{\quad} \text{ kmol } O_2}$

$\dot{V}_{air}\left(\dfrac{\text{L}}{\text{min}}\right) = \dfrac{\dot{n}_1 RT}{P} =$

$= \dfrac{\dot{n}_1(\text{kmol air})}{\text{min}} \left| \dfrac{\underline{\quad}}{\underline{\quad}} \right| \dfrac{\underline{\quad} \text{ L} \cdot \text{atm}}{\text{mol} \cdot \text{K}} \left| \dfrac{\underline{\quad} \text{ K}}{(\underline{\quad}+101.325) \text{ kPa}} \right| \dfrac{\underline{\quad} \text{ kPa}}{1 \text{ atm}}$

(5.46-4)

Unreacted pentane $\dot{n}_2(\dfrac{\text{kmol } C_5H_{12}}{\text{min}}) = \dfrac{\underline{\quad} \text{ kg } C_5H_{12}}{\text{min}} \left| \dfrac{\underline{\quad}}{\underline{\quad}} \right.$

Fractional pentane conversion $X_p = \dfrac{\underline{\quad} (\text{mol } C_5H_{12} \text{ react})}{\underline{\quad} (\text{mol } C_5H_{12} \text{ fed})}$

(5.46-5)

Carbon balance $\dot{n}_0(\text{kmol } C_5H_{12}) \times \left(\dfrac{5 \text{ kmol C}}{\text{kmol } C_5H_{12}}\right) = \dot{n}_2(5) + \textcircled{$\dot{n}_4$}(1)$

Hydrogen balance _____

Oxygen balance _____

(5.46-6)

Product gas volumetric flow rate

$$\dot{V}_{gas}\left(\frac{L}{min}\right) = \frac{\dot{n}_{gas}RT}{P} = \frac{(\underline{\hspace{2cm}})kmol}{min}\left|\frac{\underline{\hspace{1cm}}}{\underline{\hspace{1cm}}}\right|\frac{\underline{\hspace{1cm}}\ L\cdot atm}{mol\cdot K}\left|\frac{\underline{\hspace{1cm}}}{\underline{\hspace{1cm}}}\right.$$

Total condensate volume (can also be determined using the values found in the overall analysis)

$$\dot{V}_{condensate} = \frac{3.175\ kg\ C_5H_{12}(l)}{min}\left|\frac{L}{\underline{\hspace{0.5cm}}\ kg}\right. + \frac{\underline{\hspace{0.5cm}}\ (kmol\ H_2O)}{min}\left|\frac{\underline{\hspace{1cm}}}{\underline{\hspace{1cm}}}\right|\frac{\underline{\hspace{1cm}}}{\underline{\hspace{1cm}}}$$

Assume volume additivity
and neglect the influence of   Table _____
temperature on density.

(5.46-7)

Q: Is the use of the ideal gas equation of state justified for the inlet air and product gas?

A:

Air $\dfrac{RT}{P} = \dfrac{\underline{\hspace{1cm}}\ L\cdot atm}{mol\cdot K}\left|\dfrac{\underline{\hspace{2cm}}}{\underline{\hspace{2cm}}}\right|\dfrac{\underline{\hspace{2cm}}}{\underline{\hspace{2cm}}} = 9\dfrac{L}{mol}$

Since RT/P > 5 L/mol, from Eq. 5.2-3a on p. 192 of the text, the assumption of ideal gas behavior is justified.

Gas $\dfrac{RT}{P} =$

_____

Since RT/P > _____ L/mol, from Eq. 5.2-3b on p. 192 of the text, the assumption of ideal gas behavior is justified.

In the space below, write the equations as an E-Z Solve script, enter your equations into the program, and solve for all of the unknowns. Enter your answers in the space provided.

E-Z Solve Code   (5.46-8)

```
// Problem 5.46
n0 = 50.4*0.630/72.05
n1 = n0*1.15*8/0.210
Vair=
n2 =
Xp =
n0*5 = n2*5 + n4




Vgas =
Vcond =
```

E-Z Solve Solution   (5.46-9)

$\dot{n}_0 =$ _____ $\dot{n}_3 =$ _____

$\dot{n}_1 =$ _____ $\dot{n}_4 =$ _____

$\dot{V}_{air} = \underline{1.74 \times 10^5}$ L air fed / min $\dot{n}_5 =$ _____

$\dot{n}_2 =$ _____ $\dot{V}_{gas} = \underline{4.09 \times 10^5}$ L gas out / min

$X_p =$ _____ $\dot{V}_{condensate} =$ _____

(b) Sketch an apparatus that could have been used to separate the pentane and water in the condensate. *Hint:* Remember that pentane and water are immiscible.

Solution   (5.46-10)

Notes and Calculations

PROBLEM 5.62

An oxygen tank with a volume of 2.5 ft^3 is kept in a room at 50°F. An engineer has used the ideal gas equation of state to determine that if the tank is first evacuated and then charged with 35.3 lb$_m$ of pure oxygen, its rated maximum allowable working pressure (MAWP) will be attained. Operation at pressures above this value is considered unsafe.

Background Information

A photo of an oxygen tank is shown below. When full, the pressure inside the tank is about 2500 psig. The pressure in the tank shown is about 1800 psig so some of the oxygen has been used. A **pressure regulator** is used to step down the internal pressure to the desired use pressure, in this case, 200 psig or less. The regulator has two gauges; one to measure the internal pressure and one to measure the delivery pressure. The delivery pressure for the tank in the photo has been set to about 50 psig. Gas tanks such as these are usually rented—one pays for the gas plus a monthly tank rental fee and trades empty tanks for tanks that have been refilled by the gas supplier. The tanks are expensive so the gas suppliers continue to reuse them as long as possible.

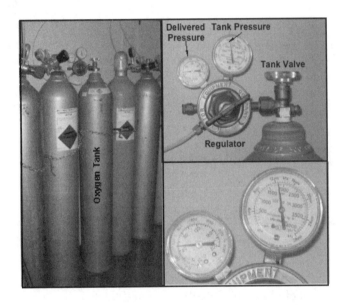

(a) What is the maximum allowable working pressure (psig) of the tank?

(b) You suspect that at the conditions of the fully charged tank, ideal gas behavior may not be a good assumption. Use the SRK equation of state to obtain a better estimate of the maximum mass of oxygen that may be charged into the tank. Did the ideal gas assumption lead to a conservative estimate (on the safe side) or a nonconservative estimate of the amount of oxygen that could be charged?

(c) Suppose the tank is charged and ruptures before the amount of oxygen calculated in part (b) enters it. (It should have been able to withstand pressures up to four times the MAWP.) Think of at least five possible explanations for the failure of the tank below its rated pressure limit.

(a) What is the maximum allowable working pressure (psig) of the tank?

Solution

From the ideal gas equation of state, the MAWP is   **(5.62-1)**

$$P_g = \frac{nRT}{V} - P_{atm}$$

$$= \frac{35.3 \text{ lb}_m \text{ O}_2}{} \left| \frac{1 \text{ lb-mol}}{\underline{\quad\quad} \text{ lb}_m} \right| \quad\quad\quad\quad | \quad\quad\quad\quad - \underline{\quad\quad} \text{ psia} = \underline{2400 \text{ psig}}$$

(b) You suspect that at the conditions of the fully charged tank, ideal gas behavior may not be a good assumption. Use the SRK equation of state to obtain a better estimate of the maximum mass of oxygen that may be charged into the tank. Did the ideal gas assumption lead to a conservative estimate (on the safe side) or a non-conservative estimate of the amount of oxygen that could be charged?

(5.62-2)

Q: Provide a basis for the suspicion that ideal gas behavior is probably not a good assumption. (*Hint*: Consult p. 192 of the text.)

A: The ideal specific molar volume of the calculated amount of oxygen in the tank is

$$\hat{V}_{ideal} = \frac{RT}{P} = \frac{10.73 \text{ ft}^3 \cdot \text{psia}}{\text{lb-mole} \cdot {}^\circ R} \left| \frac{\underline{\quad\quad} {}^\circ R}{\underline{\quad\quad} \text{ psia}} \right| = \frac{\underline{\quad\quad} \text{ ft}^3}{\underline{\quad\quad} \text{ lb-mole}}$$

Since this figure is much less than _____ ft^3/lb-mole, from Eq. _____ on p. 192, ideal gas behavior cannot be considered a good assumption.

Strategy

The SRK equation with $\hat{V}$ unknown is a cubic equation and in principle can have three different roots, only one of which is likely to be physically meaningful. E-Z Solve provides a convenient way to solve the equation, and that's the way we'll do it.

When solving a nonlinear equation using E-Z Solve or any other equation-solving program, it is necessary to estimate the solution first. The better the estimate, the more likely it is that the program will converge quickly to the desired root. We will first set up the SRK equation, substituting known quantities from the problem solution, then come up with an estimated solution, and finally use E-Z Solve to calculate the solution.

Solution

SRK Equation of state: $P = \dfrac{RT}{\left(\hat{V} - b\right)} - \dfrac{\alpha a}{\hat{V}\left(\hat{V} + b\right)}$

In work box **5.62-3** below, we've assembled all of the equations required for using the SRK equation of state. Review the material on cubic equations of state in your text beginning on p. 203 and then complete the work box by filling in the numerical values for this problem.

Name: _____

Date: _____

(5.62-3)

$T =$ _____ °R, $P =$ _____ psia

$T_c =$ _____ K = _____ °R, $P_c =$ _____ atm = _____ psi, $\omega=$ _____

$R =$ _____ $\dfrac{ft^3 \cdot psi}{lb\text{-}mole \cdot °R}$

$a =$ _____ $\dfrac{(RT_c)^2}{P_c}$, $b =$ _____ $\dfrac{RT_c}{P_c}$

$m =$ _____

$T_r = \dfrac{T}{T_c}$, $\alpha =$ _____

Initial Estimate

Use the ideal gas specific volume as a first guess ($\hat{V} =$ _____ ft^3 / lb-mole)   **(5.62-4)**

E-Z Solve Program   **(5.62-5)**

// Problem 5.62(b)

P = (R*T)/(V-b)-alpha*a/(V*(V+b))   // SRK equation, psia

a = 0.42748*((R*Tc)^2)/Pc   // ft6-psi/lbmol2

b = _____   // ft3/lbmol

m = _____

alpha = (1+m*(1–sqrt(Tr)))^2

T = _____; P = _____   // temp, R & pressure, psia

Tc = _____; Pc = _____   // crit. temp., R, & pressure, psia

w = _____   // acentric factor (Tab. 5.3-1)

R = _____   // ft3-psia/lb-mol-deg.R

Tr = T/Tc   // reduced temp

//Initial guess for V = _____ ft3/lb-mole

(5.62-6)

E-Z Solve $\Rightarrow \hat{V}=$ _____ ft^3 / lb-mole

$\Rightarrow m_{O_2} = \dfrac{2.5\ ft^3}{\text{_____ } ft^3 / lb\text{-mole}} \left| \dfrac{\text{_____ } lb_m}{lb\text{-mole}} \right. =$ _____ lb_m

(5.62-7)

The ideal gas approximation gives a ____ conservative ____ nonconservative estimate, calling for charging ____ less ____ more O_2 than the tank can safely hold.

Name: _____

Date: _____

(c) Suppose the tank is charged and ruptures before the amount of oxygen calculated in part (b) enters it. (It should have been able to withstand pressures up to four times the MAWP.) Think of at least five possible explanations for why the tank might fail at a pressure below its rated pressure limit.

(5.62-8)

- _____
- _____
- _____
- _____
- _____
- _____
- _____

PROBLEM 5.68

A 10-liter cylinder containing oxygen at 175 atm absolute is used to supply O_2 to an oxygen tent. The cylinder can be used until its absolute pressure drops to 1.1 atm. Assuming a constant temperature of 27°C, calculate the gram-moles of O_2 that can be obtained from the cylinder, using the compressibility-factor equation of state when appropriate.

Strategy

We will first do a quick estimate of whether ideal gas behavior is a reasonable approximation at the initial and final conditions in the cylinder, then use either the ideal gas equation of state or the compressibility gas factor equation of state to determine n at both conditions (in both cases, T, P, and V are known), and finally determine the moles of oxygen obtainable as the difference between the initial and final values.

Solution

(5.68-1)

Q: Without doing any calculations, speculate on whether the ideal gas equation of state is likely to be applicable at the initial and final conditions in the cylinder.

A: Initially (P=175 atm) ___ applicable ___not applicable

Finally (P=1.1 atm) ___ applicable ___not applicable

Q: Which equation in the text can be used to obtain a quick estimate of the applicability of the ideal gas EOS?

A: Eq. _____ on p. _____.

Test of ideality

$$\hat{V}_{175\,atm} = \frac{RT}{P} = \frac{\frac{L \cdot atm}{mol \cdot K} \times (\underline{\quad}) K}{\underline{\qquad} atm} = \underline{\quad}\frac{L}{mol} < \underline{\quad}\frac{L}{mol} \Rightarrow \underline{nonideal} \qquad (5.68\text{-}2)$$

$$\hat{V}_{1.1\,atm} = \qquad\qquad = \underline{\quad}\frac{L}{mol} > \underline{\quad}\frac{L}{mol} \Rightarrow \underline{\qquad} \qquad (5.68\text{-}3)$$

Compressibility factor at initial condition

$$T_r = \frac{T}{T_c} = \frac{\underline{\qquad} K}{\underline{\qquad} K} = \underline{\quad}, \quad P_r = \frac{P}{P_c} = \frac{\underline{\qquad} atm}{\underline{\qquad} atm} = \underline{\quad} \xrightarrow{\text{Fig.} \underline{\qquad}} z = 0.95 \qquad (5.68\text{-}4)$$

Equation of state

$$\textbf{Initial} \quad n_0 = \frac{P_0 V}{zRT} = \frac{\underline{\quad} atm \,\big|\, \underline{\quad} L \,\big|\, \underline{\quad} mol \cdot K}{\underline{\quad}\,\big|\,\underline{\quad} K \,\big|\, \underline{\quad} L \cdot atm} = \underline{\quad} mol\ O_2 \qquad (5.68\text{-}5)$$

$$\textbf{Final} \quad n_0 = \underline{\hspace{10cm}} = \underline{\quad} mol\ O_2 \qquad (5.68\text{-}6)$$

$$\Rightarrow \underline{\quad} mol\ O_2 \text{ obtainable} \qquad (5.68\text{-}7)$$

Notes and Calculations

PROBLEM 5.80

A gas mixture consisting of 15.0 mole% methane, 60.0% ethylene, and 25.0 mole% ethane is compressed to a pressure of 175 bar at 90°C. It flows through a process line in which the velocity should be no greater than 10 m/s. What flow rate (kmol/min) of the mixture can be handled by a 2-cm internal diameter pipe?

Strategy

The maximum velocity and internal diameter of the pipe allow us to determine a maximum volumetric flow rate (= velocity times cross-sectional area). Knowing the temperature and pressure, we can then calculate a maximum molar flow rate using an appropriate equation of state for the gas mixture.

At 175 bar, the mixture seems to be highly compressed; however, the 90°C temperature might be sufficiently high to mitigate the effect of pressure on the nonideality of the mixture. We will first apply the rule-of-thumb on p. 192 to decide whether or not to assume ideal gas behavior. If we cannot and we want to use a non-ideal equation of state from the text, we have no choice: the compressibility factor equation of state coupled with Kay's rule is the only correlation given that enables us to do *PVT* calculations for mixtures of gases.

Solution

Test of ideality

$$\hat{V} = \frac{RT}{P} = \frac{\dfrac{L \cdot bar}{mol \cdot K} \times \underline{\quad} K}{\underline{\quad} atm} = \underline{\quad} \frac{L}{mol} < \underline{\quad} \frac{L}{mol} \Rightarrow \underline{nonideal} \qquad (5.80\text{-}1)$$

Maximum volumetric flow rate

$$\dot{V}_{max} = u_{max} A_{pipe} = \frac{\underline{\quad} m}{s} \left| \frac{s}{\underline{\quad} min} \right| \frac{\underline{\quad} cm^2}{} \left| \frac{\underline{\quad} m^2}{\underline{\quad} cm^2} = \underline{\quad} \frac{m^3}{min} \qquad (5.80\text{-}2)$$

Calculation of pseudocritical temperature and pressure (Kay's rule) (5.80-3)

Critical Properties			
Component	Mol Fraction	T_c, K	P_c, atm
methane	.15	_____	_____
ethylene	.60	_____	_____
ethane	.25	_____	_____

Eq. (5.4-9) $\Rightarrow T_c' = 0.15(\underline{\quad} K) + 0.60(\underline{\quad} K) + 0.25(\underline{\quad} K) = \underline{\quad} K$

Eq. (5.4-10) $\Rightarrow P_c' = 0.15(\underline{\quad} atm) + 0.60(\underline{\quad} atm) + 0.25(\underline{\quad} atm) = \underline{\quad} atm$

Name: ____________________

Date: ____________

Calculation of pseudoreduced conditions and compressibility factor **(5.80-4)**

$$\left.\begin{array}{l} \text{Eq. (5.4-11)} \Rightarrow T_r' = \dfrac{T}{T_c'} = \dfrac{____\ \text{K}}{____\ \text{K}} = ____ \\[2ex] \text{Eq. (5.4-12)} \Rightarrow P_r' = \dfrac{P}{P_c'} = \dfrac{(____\ \text{bar})\ \dfrac{____\ \text{atm}}{____\ \text{bar}}}{____\ \text{atm}} = ____ \end{array}\right\} \xrightarrow{\text{Fig. 5.4-3}} z_m = 0.67$$

Compressibility factor equation of state

$$\dot{n}_{max} = \frac{P\dot{V}_{max}}{z_m RT} = \frac{____\ \text{bar}}{____} \left| \frac{\text{mol} \cdot \text{K}}{____\ \text{L} \cdot \text{bar}} \right| \frac{____\ \text{m}^3/\text{min}}{____\ \text{K}} = \underline{1.6\ \text{kmol/min}}$$ **(5.80-5)**

A pure species may exist as a solid, liquid, or vapor, depending on the system temperature and pressure. For example, pure water at 1 atm is a solid at temperatures below 0°C, a liquid at temperatures above 0°C and below 100°C, and a vapor above 100°C. At 0°C (the *freezing point* or *melting point* of water at 1 atm), the water may exist as a solid, a liquid, or a mixture of both, and at 100°C (the *boiling point* of water at 1 atm) the water may be liquid, vapor, or a mixture of both. At 0.0098°C and 4.58 mm Hg (the *triple point* of water), water may exist as solid, liquid, vapor, or a mixture of the three phases.

Similarly, a mixture of two or more species may be solid, liquid, vapor, or a combination of different phases (solid and liquid, liquid and vapor, or separate liquid phases), depending on the system composition, temperature, and pressure. For example, a mixture containing 50 mole% benzene and 50 mole% toluene at 1 atm is a liquid below 91°C (the *bubble point temperature* of the mixture at 1 atm) and a vapor above 99°C (the *dew point temperature* of the mixture at 1 atm). Between those two temperatures liquid and vapor phases can coexist, with the vapor containing relatively more of the more volatile of the two components (benzene).

If a mixture of species is brought to a temperature and pressure at which two phases can coexist, each species distributes itself between the phases. After enough time has elapsed, the system reaches a state of *equilibrium* at which the composition of each phase remains constant. This chapter teaches you how to calculate equilibrium compositions for a specified temperature and pressure (or how to calculate the equilibrium temperature or pressure from specified composition data) and how to incorporate the results into material balance calculations on multiphase systems at equilibrium.

PROBLEM 6.11

A gas mixture contains 10.0 mole% $H_2O(v)$ and 90.0 mole% N_2. The gas temperature and absolute pressure at the start of each of the three parts of this problem are 50°C and 500 mm Hg. Ideal gas behavior may be assumed in every part of this problem.

(a) If some of the gas mixture is put in a cylinder and slowly cooled at constant pressure, at what temperature would the first drop of liquid form?

Solution

Water is the only condensable species at the low pressure in this system. At the point where the first bubble forms, the system may be shown schematically as follows:

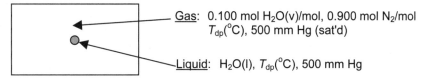

Gas: 0.100 mol $H_2O(v)$/mol, 0.900 mol N_2/mol
T_{dp}(°C), 500 mm Hg (sat'd)

Liquid: $H_2O(l)$, T_{dp}(°C), 500 mm Hg

The loss from the gas phase of the tiny amount of water that condenses to form the first droplet of liquid makes an insignificant change in the composition of the gas, which therefore remains at its initial value. (If this is not clear to you, try it with numbers. Suppose that the gas phase initially contains 1.0000 mol of water vapor and 9.0000 mol of nitrogen, then take away 0.0001 mol of the water and calculate the composition of the remaining gas.)

The relationship between the compositions of gas and liquid phases at equilibrium may be estimated using Raoult's law [Eq. (6.3–1) for a single condensable species on p. 249 in the text]:

$$P_{H_2O} = y_{H_2O}P = (\underline{})(\underline{} \text{ mm Hg}) = 50 \text{ mm Hg} = p^*_{H_2O}(T_{dp})$$

$$\xrightarrow{\text{Table B.3}} T_{dp} = \underline{} \,°C$$

(6.11-1)

(b) If a 30.0-liter flask is filled with some of the gas mixture and sealed and the water vapor in the flask is completely condensed, what volume (cm^3) would be occupied by the liquid water?

Solution

As usual when a process is described, we begin by drawing a flowchart.

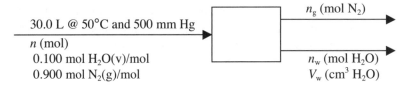

30.0 L @ 50°C and 500 mm Hg
n (mol)
0.100 mol H_2O(v)/mol
0.900 mol N_2(g)/mol

n_g (mol N_2)

n_w (mol H_2O)
V_w (cm^3 H_2O)

Ideal gas EOS

$$n = \frac{PV}{RT} = \frac{\underline{} \text{ mm Hg} \left| \underline{} \text{ L} \right.}{\underline{} \text{ K} \left| \underline{} \text{ L} \cdot \text{mm Hg}/(\text{mol} \cdot \text{K}) \right.} = 0.745 \text{ mol}$$

(6.11-2)

Water balance

$$n_w = (0.745 \text{ mol})(0.100 \, \frac{\text{mol } H_2O}{\text{mol}}) = 0.0745 \text{ mol } H_2O$$

$$\Rightarrow V_w = \frac{0.0745 \text{ mol } H_2O \left| \underline{} \text{ g} \right| \underline{} cm^3}{\left| \underline{} \text{ mol} \right| \underline{} \text{ g}} = \underline{} cm^3$$

(6.11-3)

(c) If the gas mixture is stored in a rigid-walled cylinder and a low-pressure weather front moves in and the barometric (atmospheric) pressure drops, which of the following would change: (i) the gas density, (ii) the absolute pressure of the gas, (iii) the partial pressure of water in the gas, (iv) the gauge pressure of the gas, (v) the mole fraction of water in the gas, (vi) the dew-point temperature of the mixture? Explain your answer.

Solution

(6.11-4)

PROBLEM 6.18

Air at 90°C and 1.00 atm (absolute) contains 10.0 mole% water. A continuous stream of this air enters a compressor-condenser, in which the temperature is lowered to 15.6°C and the pressure is raised to 3.00 atm. The air leaving the condenser is then heated isobarically to 100°C. Calculate the fraction of water that is condensed from the air, the relative humidity of the air at 100°C, and the ratio m^3 outlet air @ 100°C /m^3 feed air @ 90°C.

Solution

Basis: 1 mol feed. Since the problem statement asks us to calculate the ratio of the volumes of feed air and exit air, we will label both volumes on the flowchart.

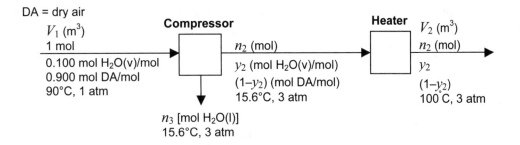

DEGREE-OF-FREEDOM ANALYSIS ON COMPRESSOR		
UNKNOWNS AND INFORMATION		**JUSTIFICATION/CONCLUSION**
+ 4 unknowns	(V_1, n_1, n_2, y_2)	
− 2 balances	_____ , _____	
− 1 gas law at inlet		
1 DOF		Problem is unsolvable

Since we have more unknowns than equations, unless we can come up with another relationship among the compressor variables we're stuck. (If you do the DOF analyses for the overall system and the heater you'll run into the same problem—try it.) Fortunately, there is another relationship. Can you state what it is and justify your claim? (*Hint*: What do you know about the two streams leaving the compressor?)

(6.18-2)

The solution strategy is straightforward.

Name: _____

Date: _____

(6.18-3)

- Raoult's law applied to the compressor outlet stream $[\,y_2 P = p^*_{H_2O}(\rule{1.5cm}{0.15mm}\,°C)\,]$ contains only _____ as an unknown, and so we can begin by solving for _____.
- A mole balance on _____ can then be solved for n_2 and a _____ balance can be solved for n_3.
- Since we now know how much water is entering (_____ mol) and how much is condensed (_____), we can calculate the fractional condensation.
- Since we know the temperature, pressure, and mole fraction of water in the heater exit air, we can calculate the relative humidity of the air.
- Finally, since we know the molar flow rates of the feed and exit air and the temperatures and pressures of both streams, we can use the ideal gas equation of state to determine both V_1 and V_2, and hence the ratio V_2/V_1.

All that remains is the algebra.

Saturation

$$y_2 = \frac{p^*_{H_2O}\left(\rule{1.2cm}{0.15mm}\,°C\right)}{P} \xRightarrow{\text{Table B.3}} y_2 = \frac{\rule{1.5cm}{0.15mm}\ \text{mm Hg}}{\rule{1.2cm}{0.15mm}\ \text{atm}}\ \left|\ \frac{1\ \text{atm}}{760\ \text{mm Hg}}\right.$$

$$= 0.00583\ \text{mol}\ H_2O(v)/\text{mol}$$

(6.18-4)

Dry air balance

$$0.900(1) = \rule{4cm}{0.15mm} \Rightarrow n_2 = 0.9053\ \text{mol}$$

(6.18-5)

H₂O balance

$$\rule{2cm}{0.15mm} = \rule{5cm}{0.15mm} \Rightarrow n_3 = 0.0947\ \text{mol}$$

(6.18-6)

Fraction H₂O condensed

$$\text{fraction} = \frac{\rule{2cm}{0.15mm}\ \text{mol condensed}}{\rule{2cm}{0.15mm}\ \text{mol fed}}$$

$$= \rule{1.5cm}{0.15mm}\ \text{mol}\ H_2O\,\text{condense}/\text{mol}\ H_2O\ \text{fed}$$

(6.18-7)

Definition of relative saturation (or relative humidity), Eq. (6.3-4), p. 253 in the text:

$$h_r = \frac{y_2 P_{outlet}}{p^*_{H_2O}\left(100°C\right)} \times 100\% = \frac{\rule{1.5cm}{0.15mm}\left(\rule{1cm}{0.15mm}\ \text{atm}\right)}{\rule{1cm}{0.15mm}\ \text{atm}} \times 100\% = \rule{1.5cm}{0.15mm}\ \%$$

(6.18-8)

Ideal gas law for V_1 and V_2:

$$V_2 = \frac{0.9053 \text{ mol} \left| \underline{\hspace{1cm}} \text{L (STP)} \right| \underline{\hspace{0.5cm}} \text{K} \left| 1 \text{ atm} \right| 1 \text{ m}^3}{\left| \text{mol} \right| 273\text{K} \left| \underline{\hspace{0.5cm}} \text{atm} \right| 10^3 \text{ L}}$$ (6.18-9)

$$= 9.24 \times 10^{-3} \text{m}^3 \text{ outlet air @ } 100°\text{C}$$

$$V_1 = \frac{1 \text{ mol} \left| \underline{\hspace{1cm}} \text{L (STP)} \right| \underline{\hspace{0.5cm}} \text{K} \left| 1 \text{ m}^3 \right.}{\left| \text{mol} \right| 273\text{K} \left| 10^3 \text{ L} \right.} = \underline{\hspace{1.5cm}} \text{ m}^3 \text{ feed air @ } 90°\text{C}$$ (6.18-10)

$$\frac{V_2}{V_1} = \frac{\underline{\hspace{1.5cm}} \text{ m}^3 \text{ outlet air}}{\underline{\hspace{1.5cm}} \text{ m}^3 \text{ feed air}} = \underline{\hspace{1.5cm}} \text{ m}^3 \text{ outlet air}/\text{m}^3 \text{ feed air}$$ (6.18-11)

Notes and Calculations

PROBLEM 6.24 (Modified)

A 20,000-liter storage tank was taken out of service to repair and reattach a feed line damaged in a collision with a tanker. The tank was drained and then opened several days later for a welder to enter and perform the required work. No one realized, however, that 15 liters of liquid nonane (C_9H_{20}) remained in a collection sump at the bottom of the tank after the draining had been completed.

(a) Nonane has a lower explosion limit of 0.80 mole% and an upper explosion limit of 2.9 mole% (i.e., nonane-air mixtures at 1 atm can explode when exposed to a spark or flame if the nonane mole percentage is between the two given values). Assume any liquid nonane that evaporates spreads uniformly throughout the tank. Is it possible for the average gas-phase composition in the tank to be within the explosion limits at any time? If the answer is yes, why might the actual percentage of nonane in the tank gas always be below the lower explosion limit? If the answer is no, why might explosion still be a possibility?

Strategy

The greatest possible mole fraction of nonane in the tank gas (assuming the nonane concentration is equal everywhere) would be attained if all the nonane in the sump evaporated and none escaped.

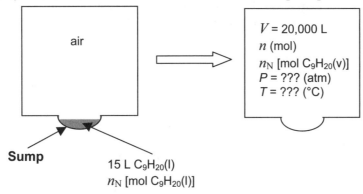

(6.24-1)

- Calculate n_N from the given liquid volume and _____ (from Table B.1).

- Estimate n from _____, assuming a reasonable value for T.

- Calculate the maximum mole% of nonane in the tank as $p_N = (n_N/n) \times 100\%$.

- If $p_N < 0.80\%$, then the average nonane composition in the tank cannot fall within the explosion limits. If $p_N \geq 0.80\%$, then the average nonane composition in the tank can fall within the explosion limits, either at steady state (if $p_N \leq 2.9\%$) or while the liquid nonane is evaporating (if $p_N > 2.9\%$). (We'll deal with the rest of the question later.)

Solution

Moles of nonane: We assume that after the tank is drained, the only nonane it contains is the liquid in the sump; that is, we neglect any nonane that may initially be in the vapor phase.

Table B.1

$$n_N = \frac{15 \text{ L } C_9H_{20}(l)}{} \left| \frac{\times 1.00 \text{ kg}}{\text{L } C_9H_{20}} \right| = 0.084 \text{ kmol } C_9H_{20} \qquad (6.24\text{-}2)$$

Total moles of gas in the tank and nonane percentage: Assume ideal gas behavior and an ambient temperature of 25°C and a pressure of 1 atm. (The conclusions will not be significantly affected by slight variations in the assumed values.)

$$n = \frac{2 \times 10^4 \text{ L}}{} \left| \frac{\text{\_\_\_\_ K}}{\text{\_\_\_\_ K}} \right| \frac{1}{} \left| \frac{\text{kmol}}{\text{_____ L(STP)}} \right| = \text{\_\_\_\_\_ kmol}$$

(6.24-3)

$$\Rightarrow p_N = \frac{n_N}{n} \times 100\% = \frac{\text{\_\_\_\_\_ mol } C_9H_{10}}{\text{\_\_\_\_\_ mol}} \times 100\% = 10\% \text{ nonane}$$

(6.24-4)

Q: Is it possible for the average gas phase composition in the tank (that is, the total nonane in the tank divided by the total moles of gas) to be within the explosion limits at any time?

A: _____

_____

_____

Q: Despite the result just given, why might the actual average composition in the tank always be below the lower explosion limit? (Think about the assumptions we made to obtain that result.)

A: _____

_____

_____

Q: Even if the maximum mole percent of nonane were below the lower explosive limit, why might explosion still be a danger? (Again, think about the assumptions.)

A: _____

_____

_____

_____

(b) Nonane has a vapor pressure of 5.00 mm Hg at 25.8°C and 40.0 mm Hg at 66°C. Use the Clausius-Clapeyron equation (Eq. 6.1-3, p. 244 in the text) to derive an expression for $p^*(T)$. Then, calculate the temperature at which the system would have to equilibrate in order for the gas in the tank to be at the lower explosion limit.

Strategy

There are two possibilities for equilibration in the tank. (1) All of the nonane liquid evaporates and none of the nonane vapor escapes, and (2) some of the liquid evaporates, none of the vapor escapes, and the gas in the tank becomes saturated with nonane vapor at the system temperature. In the first of these cases, the gas in the tank would be well above the upper explosion limit [as we showed in Part (a)] and its composition would not depend on the temperature. The only way the problem can be meaningful is if the second case is applicable, and so we will assume that it is. The system at equilibrium with the gas composition at the lower explosion limit appears as shown in the following diagram.

C_9H_{20} (l)

0.008 mol C_9H_{20}(v)/mol, sat'd
0.992 mol air/mol
T (°C), 1 atm

- Raoult's law (Eq. 6.3-1) relates $y^*_{C_9H_{20}}(= 0.008)$, $P (= 1 \text{ atm})$, and $p^*_{C_9H_{20}}(T)$. Since the first two quantities are known, we can calculate the third one.
- The Clausius-Clapeyron equation, which relates the vapor pressure to the temperature, has two parameters (A and B) and we are given two data points for $p^*_{C_9H_{20}}$ vs. T, so we can determine A and B algebraically. Once we have done so, we can estimate $p^*_{C_9H_{20}}$ for a given T or—as we are required to do in this problem—estimate T for a known $p^*_{C_9H_{20}}$. We will begin with the second of these steps and then do the first one.

Solution
Clausius-Clapeyron equation

$$\ln p^* = -\frac{A}{T} + B$$

$\downarrow$ $T_1 = 25.8°C = 299 \text{ K}, p^*_1 = 5.00 \text{ mmHg}; \; T_2 = 66.0°C = 339 \text{ K}, p^*_2 = 40.0 \text{ mmHg}$

$$-A = \frac{\ln(p^*_2 / p^*_1)}{\dfrac{1}{T_2} - \dfrac{1}{T_1}} = \frac{\ln(\underline{\quad} / \underline{\quad})}{\dfrac{1}{\underline{\quad}} - \dfrac{1}{\underline{\quad}}} \Rightarrow A = 5269$$

$$B = \ln(p^*_1) + \frac{A}{\underline{\quad}} = \ln(\underline{\quad}) + \overline{\underline{\quad}} = 19.23 \tag{6.24-4}$$

$$\Rightarrow \ln p^* = 19.23 - \frac{5269}{T(\text{K})} \Rightarrow p^* = \underline{\hspace{4cm}}$$

Raoult's Law (Eq. 6.3-1)

$$y_N P = p^*_N(T) \Rightarrow \underline{\quad}(\underline{\quad} \text{ mm Hg}) = \underline{\quad} \text{ mm Hg}$$

$$\xrightarrow{\text{C-C equation}} \ln(\underline{\quad}) = 19.23 - \frac{5269}{T} \Rightarrow T = 302 \text{ K} = \underline{29°C} \tag{6.24-5}$$

(c) Fortunately, a safety inspector examined the system before the welder began work and immediately cancelled the work order. The welder was cited and fined for violating established safety procedures. One requirement was for the tank to be purged thoroughly with steam after being drained.

(6.24-6)

Q: What is the purpose of this requirement? (Why purge, and why with steam rather than air?)
A: The purpose of purging is to _____ _____
Using steam rather than air ensures _____ _____
Q: What other precautions should be taken to be sure that the welder is in no danger?
A: Before anyone goes into the tank, _____ _____

Notes and Calculations

Name: _____

Date: _____

PROBLEM 6.32

A gas stream containing 40.0 mole% hydrogen, 35.0% carbon monoxide, 20.0% carbon dioxide, and 5.0% methane is cooled from 1000°C to 10°C at a constant absolute pressure of 35.0 atm. Gas enters the cooler at 120 m³/min and upon leaving the cooler is fed to an absorber, where it is contacted with refrigerated liquid methanol. The methanol is fed to the absorber at a molar flow rate 1.2 times that of the inlet gas and absorbs essentially all of the CO_2, 98% of the methane, and none of the other components of the feed gas. The gas leaving the absorber, which is saturated with methanol at −12°C, is fed to a cross-country pipeline.

(a) Calculate the volumetric flow rate of the gas stream entering the absorber (m³/min) and the molar flow rate of methanol in the gas leaving the absorber. *Do not assume ideal gas behavior when doing PVT calculations.*

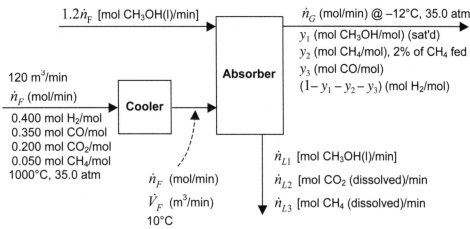

Strategy

- We will first do a degree-of-freedom analysis on the overall system to make sure we have enough information to determine the requested quantities ($\dot{V}_F, \dot{n}_G, y_1$).
- Assuming we do, we will then use an equation of state to convert the volumetric flow rate of the feed stream (120 m³/min) to a molar flow rate and the latter to the volumetric flow rate of the cooler outlet stream ($\dot{V}_F$). Since each stream is a mixture of species and Chapter 5 only presents one way to do PVT calculations for mixtures (the compressibility factor equation of state with Kay's rule), we'll use that one.
- Finally, we'll write and solve the equations listed in the DOF analysis.

Solution

(6.32-1)

DEGREE-OF-FREEDOM ANALYSIS: OVERALL SYSTEM		
UNKNOWNS AND INFORMATION		**JUSTIFICATION/ CONCLUSION**
+ 8 unknowns	_____	
− 5 material balances	_____	
− 1 eq. of state at cooler inlet	Calculate _____	
− 1 Raoult's law for CH_3OH	Calculate _____	
− 1 98% CH_4 absorption		_____
0 DOF		Problem is solvable

- Once we know $\dot{n}_F$, we can use an equation of state to calculate the volumetric flow rate of the cooler outlet stream ($\dot{V}_F$).

- We will use Kay's rule [Eqs. (5.4-9)–(5.4-12), p. 211 in the text] to determine the pseudoreduced temperatures and pressures of the cooler feed and outlet gases. The compressibility charts will be used to calculate the mixture compressibility factors for both streams, and the compressibility factor equation of state (Eq. 5.4-13) will be used to calculate (1) the streams' molar flow rate ($\dot{n}_F$) from the volumetric flow rate of the feed stream (120 m^3/min) and (2) the volumetric flow rate of the cooler outlet gas ($\dot{V}_F$) from $\dot{n}_F$.

Critical constants for feed gas species   (6.32-2)

Cmpd.	$T_c(K)$	$P_c(atm)$	$(T_c)_{corr}$	$(P_c)_{corr}$	
H_2	_____	_____	_____	_____	$(T_c$ & P_c from Table B.1)
CO	133.0	34.5	–	–	Newton's corrections: [Eq. (5.4-4) & (5.4-5), p. 208]
CO_2	304.2	72.9	–	–	
CH_4	190.7	45.8	–	–	

Kay's rule and compressibility factors [Eqs. (5.4-9)–(5.4-12)]   (6.32-3)

$$T'_c = \sum y_i T_{ci} = 0.40(41.3) + 0.35(133.0) + 0.20(304.2) + 0.05(190.7) = 133.4K$$

$$P'_c = \sum y_i P_{ci} = 0.40(20.8) + 0.35(34.5) + 0.20(72.9) + 0.05(45.8) = 37.3 \text{ atm}$$

Cooler inlet
$$\begin{aligned} T_r &= \underline{\quad} K/\underline{\quad} K = 9.54 \\ P_r &= \underline{\quad} atm/\underline{\quad} atm = 0.94 \end{aligned} \Bigg\} \text{Fig.} \underline{\qquad} \rightarrow z_{in} = \underline{\quad}$$

Cooler outlet
$$\begin{aligned} T_r &= \underline{\quad} K/\underline{\quad} K = 2.12 \\ P_r &= \underline{\quad} atm/\underline{\quad} atm = 0.94 \end{aligned} \Bigg\} \text{Fig.} \underline{\qquad} \rightarrow z_{out} = \underline{\quad}$$

Compressibility factor equation of state (5.4-13)—cooler inlet   (6.32-4)

$$\hat{V}_{in} = \frac{z_{in}RT}{P} = \underline{\quad} = \frac{\underline{\quad}}{\underline{\quad} \text{ atm}} \left| \frac{\underline{\quad} \text{ N}\cdot\text{m}}{\text{mol}\cdot\text{K}} \right| \underline{\quad} K \left| \frac{1 \text{ atm}}{\underline{\quad} \text{ N/m}^2} \right. = 3.04 \times 10^{-3} \frac{m^3}{\text{mol}}$$

$$\Rightarrow \dot{n}_F = \frac{\underline{\quad} \text{ m}^3}{\text{min}} \left| \frac{\text{mol}}{\underline{\quad} \text{ m}^3} \right. = 3.95 \times 10^4 \text{ mol/min}$$

Compressibility factor equation of state—cooler outlet   (6.32-5)

$$\hat{V}_{out} = \frac{z_{out}RT}{P} = \underline{\quad} = \frac{\underline{\quad}}{\underline{\quad}} \left| \frac{\underline{\quad} \text{ N}\cdot\text{m}}{\text{mol}\cdot\text{K}} \right| \underline{\quad} \left| \frac{1 \text{ atm}}{\underline{\quad} \text{ N/m}^2} \right. = \underline{\quad} \frac{m^3}{\text{mol}}$$

$$\Rightarrow \dot{V}_F = \frac{\underline{\quad} \text{ mol}}{\text{min}} \left| \frac{\underline{\quad} \text{ m}^3}{\text{mol}} \right. = \underline{\quad} m^3/\text{min}$$

Raoult's law for methanol (Eq. 6.3-1, p. 249 in the text)

Table _____

$$y_1 = \frac{p^*_{CH_3OH}\left(\underline{\quad\quad} K\right)}{P} = \frac{\left[10^{7.87863 - \underline{\quad\quad}/\left(-12 + \underline{\quad\quad}\right)}\right] mm\ Hg}{\underline{\quad\quad} atm \times \frac{760\ mm\ Hg}{atm}} \tag{6.32-6}$$

$$= 4.97 \times 10^{-4}\ mol\ CH_3OH/mol$$

98% CH$_4$ absorption (2% not absorbed)

$$(\dot{n}_G y_2) = 0.02(\underline{\hspace{4cm}})\frac{mol\ CH_4}{min} \tag{6.32-7}$$

CO balance

$$\underline{\hspace{3cm}}\frac{mol\ CO}{min} = \dot{n}_G y_3 \tag{6.32-8}$$

H$_2$ balance

$$\underline{\hspace{4cm}} = \underline{\hspace{4cm}} \tag{6.32-9}$$

We can solve the last three equations in three unknowns (use E-Z Solve) to calculate

$$\dot{n}_G = \underline{\hspace{3cm}} mol/min$$

$$\Rightarrow \dot{n}_{CH_3OH} = \dot{n}_G y_1 = \underline{\hspace{3cm}} mol\ CH_3OH/min \tag{6.32-10}$$

The values of y_2 and y_3 are not needed for the problem solution, nor is it necessary to write methanol, CH$_4$, and CO$_2$ balances to determine the three liquid outlet stream component molar flow rates.

(b)

$$\tag{6.32-11}$$

> **Q:** What is a possible intended use of the product gas? Why is it desirable to remove the CO$_2$ from the gas prior to feeding it to the pipeline.
>
> **A:** _____
>
> _____
>
> _____

Notes and Calculations

PROBLEM 6.39
A mixture of propane and butane is burned with air. Partial analysis of the stack gas produces the following dry-basis volume percentages: 0.0527% C_3H_8, 0.0527% C_4H_{10}, 1.48% CO, and 7.12% CO_2. The stack gas is at an absolute pressure of 780 mm Hg and the dew point of the gas is 46.5°C. Calculate the molar composition of the fuel.

Solution
Basis 100 mol dry stack gas

$$C_3H_8 + 5O_2 \rightarrow 3CO_2 + 4H_2O \qquad C_4H_{10} + \frac{13}{2}O_2 \rightarrow 4CO_2 + 5H_2O$$

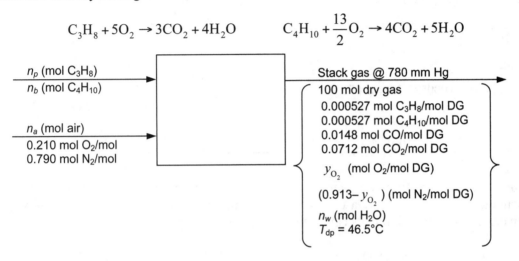

n_p (mol C_3H_8)
n_b (mol C_4H_{10})

n_a (mol air)
0.210 mol O_2/mol
0.790 mol N_2/mol

Stack gas @ 780 mm Hg
100 mol dry gas
0.000527 mol C_3H_8/mol DG
0.000527 mol C_4H_{10}/mol DG
0.0148 mol CO/mol DG
0.0712 mol CO_2/mol DG
y_{O_2} (mol O_2/mol DG)
$(0.913 - y_{O_2})$ (mol N_2/mol DG)
n_w (mol H_2O)
T_{dp} = 46.5°C

Notice that oxygen and nitrogen must be present in the stack gas even though the problem statement does not mention them, and so they are shown on the flowchart. We will base the degree-of-freedom analysis on atomic species balances.

(6.39-1)

DEGREE-OF-FREEDOM ANALYSIS		
UNKNOWNS AND INFORMATION		**JUSTIFICATION/CONCLUSION**
+ __ unknowns	_____	
− __ atomic species balances	_____	Reactive substances
− __ molecular species balance	_____	Nonreactive substances
− 1 _____		
0 DOF		Problem is solvable

Strategy
- Pure butane at 1 atm condenses at –0.6°C (from Table B.1) and pure propane condenses at an even lower temperature. Clearly, the only substance in the stack gas that would condense at 46.5°C and 1 atm (or close to it) is water. We will therefore treat water as the only condensable substance in the stack gas and apply Raoult's law for the definition of the dew point (Eq. 6.3-3) to obtain a value for the mole fraction of water in the stack gas. Since the mole fraction is (by definition) $n_w/(100+n_w)$, we will then be able to calculate n_w.

- Mole balances on carbon and hydrogen will yield the values of n_p and n_b, and once we know them we can calculate the molar composition of the fuel. Since that's all the problem asks for, we can stop right there rather than proceeding to the remaining balances.

Solution. Set up the equations but don't do the arithmetic yet.

Dew point of stack gas:

Table _____

$$T_{dp} = 46.5°C \Rightarrow y_w P = p_w^*(46.5°C) = \underline{\hspace{2cm}} \text{ mm Hg} \Rightarrow y_w = \frac{\underline{\hspace{1.5cm}} \text{ mm Hg}}{\underline{\hspace{1.5cm}} \text{ mm Hg}} \qquad \textbf{(6.39-2)}$$

Mole fraction of water (by definition):

$$y_w = \frac{n_w}{100 + n_w} \Rightarrow n_w = \frac{\overline{\underline{\hspace{1.5cm}}}}{\underline{\hspace{1.5cm}}} \qquad \textbf{(6.39-3)}$$

C balance:

$$3n_p + 4n_b = 100[0.000527(3) + (0.000527(4) + 0.0148 + 0.0712] \qquad \textbf{(6.39-4)}$$

H balance:

$$\underline{\hspace{4cm}} = (100)\left[\underline{\hspace{4cm}}\right] + \underline{\hspace{0.7cm}} n_w \qquad \textbf{(6.39-5)}$$

It would of course be straightforward to solve these four equations (the last two simultaneously) with a calculator, but to illustrate a point we suggest that you solve them using E-Z Solve, which retains many digits of precision in its calculations. The solutions are

$$y_w = \underline{\hspace{3cm}} \frac{\text{mol H}_2\text{O}}{\text{mol}}, \ n_w = 11.0478 \text{ mol H}_2\text{O},$$

$$n_p = \underline{\hspace{2.5cm}} \text{ mol C}_3\text{H}_8, \ n_b = \underline{\hspace{2.5cm}} \text{ mol C}_4\text{H}_{10} \qquad \textbf{(6.39-6)}$$

$\Rightarrow$ The fuel contains 49.1% propane and 50.9% butane

The set of four equations from **(6.39-2)** to **(6.39-5)** is *ill-conditioned*, which means that slight round-off errors in intermediate calculations can lead to substantial variations in the results. Suppose we used a calculator and rounded off to three significant digits with each step. Here is what we would get:

$y_w = 0.0995, \ n_w = 99.5/(1-.0995) = 11.0 \Rightarrow n_p = 1.53, \ n_b = 1.93 \Rightarrow$ 44% propane and 56% butane

which is considerably different from the more precise solution of **(6.39-6)**.

It is not trivial and usually more trouble than it is worth to test every set of equations you solve to see if they are ill-conditioned. A more practical approach is to retain full precision in all of your intermediate calculations. Using E-Z Solve (or any other equation-solving program) is a good way to do that.

PROBLEM 6.54

A liquid mixture containing 50.0 mole% propane, 30.0% *n*-butane, and 20.0% isobutane is stored in a rigid container at 77°F. The container has a maximum allowable working pressure of 200 psig. The head space above the liquid contains only vapors of the three hydrocarbons.

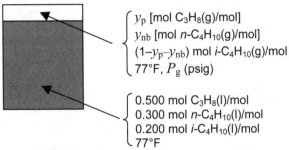

$$y_p \text{ [mol } C_3H_8(g)/mol]$$
$$y_{nb} \text{ [mol } n\text{-}C_4H_{10}(g)/mol]$$
$$(1-y_p-y_{nb}) \text{ mol } i\text{-}C_4H_{10}(g)/mol$$
$$77°F, P_g \text{ (psig)}$$

0.500 mol C_3H_8(l)/mol
0.300 mol n-C_4H_{10}(l)/mol
0.200 mol i-C_4H_{10}(l)/mol
77°F

(a) Show that the container is currently safe.
(b) Estimate the temperature above which the maximum allowable pressure would be exceeded. Comment on the suitability of the container to store the given mixture.

Strategy

(6.54-1)

Q: Which of the multicomponent vapor-liquid equilibrium correlations presented in the text—Raoult's law or Henry's law—is appropriate to use for this system? Explain your reasoning.

A: _____

_____

_____

Raoult's law: (Eq. (6.4-1) on p. 257 in the text)

$$p_P = x_p p_p^*(T) = 0.500 p_P^*(77°F)$$

p_{nB} = _____   (6.54-2)

p_{iB} = _____

$\Rightarrow P = p_P + p_{nB} + p_{iB} =$ _____

(6.54-3)

Q: To determine the pressure in the tank, we need only determine and insert the three specified vapor pressures in Eqs. **(6.54-2)**. What are the two sources of hydrocarbon vapor pressure data in the text, and which one can be used for this problem?

A: The Antoine equation (Table B.4) and the Cox chart (Fig. 6.1-4 on p. 247 of the text). The only one we can use for this problem is _____ because _____

_____

_____

(a) Show that the container is currently safe.

Solution

We will use Eqs. **(6.54-2)** to calculate the partial pressures of the three species at 77°F, add them to calculate the total pressure, and see if it is below the maximum allowable working pressure.

Cox chart (Using Figure 6.1-4 on p. 247 in the text)

$$p_P^* \approx 140 \text{ psi}, \; p_{nB}^* \approx \text{\_\_\_\_ psi}, \; p_{iB}^* \approx \text{\_\_\_\_ psi} \qquad \text{(6.54-4)}$$

Assume atmospheric pressure is 1 atm (to convert absolute to gauge pressure). From Eqs. **(6.54-2)**,

$$\text{(6.54-5)}$$

$p_P = 0.500(140 \text{ psi}) = 70 \text{ psi}, \; p_{nB} = 0.300(\text{\_\_\_\_ psi}) = \text{\_\_\_\_ psi}, \; p_{iB} = 0.200(\text{\_\_\_ psi}) = \text{\_\_\_ psi}$

$P = (70 + \text{\_\_\_\_} + \text{\_\_\_\_}) \text{ psi} = \text{\_\_\_\_ psia} - \text{\_\_\_\_\_ psia} = \text{\_\_\_\_ psig} < 200 \text{ psig} \Rightarrow \underline{\text{Currently safe.}}$

$$\text{(6.54-6)}$$

Q: The Cox chart is difficult to read with accuracy, and those three vapor pressures could be off by 5–10 psi. Is that a problem?

A: Not in this case, because _____

_____

_____

$$\text{(6.54-7)}$$

Q: What molar percentage of the gas in the head space is propane?

A: _____

_____

(b) Estimate the temperature above which the maximum allowable pressure would be exceeded. Comment on the suitability of the container to store the given mixture.

Solution

This problem must be solved by trial-and-error. A temperature is assumed, the vapor pressures of the three species are found on the Cox chart and substituted into Eqs. **(6.54-2)** to calculate the total pressure, and a new temperature is tried until the calculated P equals about 215 psia (= 200 psig).

$$= 0.500\,p_P^*(T) + 0.300\,p_{nB}^*(T) + 0.200\,p_{iB}^*(T)$$

$$\text{(6.54-8)}$$

$T(°F)$	p_P^* (psi)	p_{nB}^* (psi)	p_{iB}^* (psi)	P(psi)
130	___	___	___	___
150	___	___	___	___
140	___	___	___	___

The value at 140°F is close enough to 215, considering the accuracy with which we can read the chart.

(6.54-9)

> **Q:** Within what temperature range is the pressure likely to reach its maximum allowable value?
>
> **A:** _____

Note: If we had Antoine constants for all three species, the trial-and-error calculation could have been done much more accurately and conveniently. The Antoine expressions for $p^*(T)$ could have been entered into E-Z Solve and added to obtain a nonlinear expression for $P(T)$. A value for P of 214.7 could also have been entered, and the program would then solve for T.

Notes and Calculations

PROBLEM 6.60

The feed to a distillation column is a 45.0 mole% *n*-pentane – 55.0 mole% *n*-hexane liquid mixture. The vapor stream leaving the top of the column, which contains 98.0 mole% pentane and the balance hexane, goes to a total condenser (one in which all the vapor is condensed). Half of the liquid condensate is returned to the top of the column as *reflux* and the rest is withdrawn as overhead product (*distillate*) at a rate of 85.0 kmol/h. The distillate contains 95.0% of the pentane fed to the column. The liquid stream leaving the bottom of the column goes to a *reboiler*. Part of the stream is vaporized; the vapor is recycled to the bottom of the column as *boilup,* and the residual liquid is withdrawn as bottoms product.

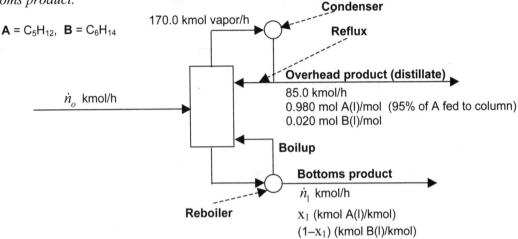

(a) Calculate the molar flow rate of the feed stream and the molar flow rate and composition of the bottoms product stream.

Solution

(6.60-1

DEGREE-OF-FREEDOM ANALYSIS	
UNKNOWNS AND INFORMATION	**JUSTIFICATION/CONCLUSION**
+ ___ unknowns _____	
– ___ balances _____	
– 1 (95% of A fed goes to distillate)	
0 DOF	Problem is solvable

95% of A fed goes to distillate

$$\left(85.0\ \frac{\text{mol}}{\text{h}}\right)\left(0.980\ \frac{\text{mol A}}{\text{mol}}\right) = 0.950(\underline{\hspace{2cm}}) \Rightarrow \dot{n}_0 = 195\ \text{kmol/h} \qquad \text{(6.60-2)}$$

Total mole balance

$$\underline{\hspace{5cm}} \Rightarrow \dot{n}_1 = \underline{\hspace{3cm}}\ \text{kmol/h} \qquad \text{(6.60-3)}$$

Pentane balance

$$\underline{\hspace{6cm}} \Rightarrow x_1 = \underline{0.0405}\ \text{mol A/mol} \qquad \text{(6.60-4)}$$

Name: _____

Date: _____

(b) Estimate the temperature of the vapor entering the condenser, assuming that it is saturated (at its dew point) at an absolute pressure of 1 atm and that Raoult's law applies to both pentane and hexane. Then estimate the volumetric flow rates of the vapor stream leaving the column and of the liquid distillate product. State any assumptions you make.

Solution

(6.60-5)

> **Q:** Suppose the problem statement had not told you to use Raoult's law. What statement in the text would have led you to conclude that Raoult's law might be applicable?
>
> **A:** _____
>
> _____

The dew point temperature of the overhead vapor stream can be calculated from Eq. (6.4-7) on p. 260 in the text, substituting the known values of the pressure (760 mm Hg) and the pentane and hexane mole fractions (y_A and y_B).

$$\frac{y_A P}{p_A^*(T_{dp})} + \frac{y_B P}{p_B^*(T_{dp})} = 1 \xrightarrow{\text{Table B.4}} \frac{0.980(760)}{10^{\,6.84471-\frac{1060.793}{T_{dp}+231.541}}} + \frac{(\underline{\quad})}{10^{\,\underline{\quad}-\frac{\underline{\quad}}{T_{dp}+\underline{\quad}}}} = 1 \qquad \textbf{(6.60-6)}$$

Here are three possible ways to solve Eq. **(6.60-6)** for T_{dp}.

- **Trial-and-error with a calculator.** Substitute values of T_{dp} into Eq. **(6.60-6)** and find the one for which the left-hand side equals 1. Use this method only if you have more spare time than you know what to do with.

- **Spreadsheet.** Enter a value of T_{dp} into one cell, and a function that calculates the left-hand side of Eq. **(6.60-6)** in another cell. Use the goal-seek function to find the value in the first cell that leads to a value of 1 in the second cell.

- **E-Z Solve.** An E-Z Solve program to solve Eq. **(6.60-6)** follows. Complete it.   **(6.60-7)**

```
xa = 0.980*760/10^(6.84471–1060.793/(Tdp+231.541))

xb = _____

xa + xb = _____
```

Enter this program into E-Z Solve, and calculate the values of x_A, x_B and T_{dp}. Complete **(6.60-8)** below and save the program for further use in Part (c).

Solution

$$x_A = 0.940 \ \frac{\text{mol A}}{\text{mol}}, \ x_B = 0.040 \ \frac{\text{mol B}}{\text{mol}}, \ T_{dp} = \underline{\quad\quad} \qquad \textbf{(6.60-8)}$$

Name: _____

Date: _____

> **Q:** We could have entered Eq. **(6.60-6)** in one line in E-Z Solve, but instead chose to calculate the two terms on the left-hand side separately (which we called x_A and x_B) and then add them. We did so partly because it is good practice not to make lines of computer code too long (breaking them up makes it easier to debug the program if something goes wrong), but also because x_A and x_B have meaning. What is their physical significance? (*Hint:* Review the derivation of Eq. 6.4-7 on p. 260 in the text.)
>
> **A:** _____
>
> _____
>
> _____

Flow rate of column overhead vapor effluent:

Assuming ideal gas behavior,

$$\dot{V}_{vapor} = \frac{\dot{n}RT}{P} = \underline{\quad} \frac{kmol}{h} \left| \underline{\quad} \frac{m^3 \cdot atm}{kmol \cdot K} \right| \frac{K}{\underline{\quad} atm} = \underline{\underline{4330 \ m^3/h}}$$ (6.60-10)

Flow rate of liquid distillate product:

Table B.1 $\Rightarrow \rho_A = 0.621$ kg/L, $\rho_B = \underline{\qquad}$ kg/L   (6.60-11)

Assuming volume additivity (see p. 151 of the text),

$$\dot{V}_{distillate} = \frac{0.98(85) \ kmol \ A}{h} \left| \frac{72.15 \ kg \ A}{kmol \ A} \right| \frac{L}{0.621 \ kg \ A}$$

$$+ \frac{\underline{\quad} \ kmol \ B}{h} \left| \frac{\underline{\quad} \ kg \ B}{kmol \ B} \right| \frac{L}{\underline{\quad} \ kg \ B} = \underline{\underline{\quad L/h}}$$ (6.60-12)

(c) Estimate the temperature of the reboiler and the composition of the vapor boilup, again assuming operation at 1 atm.

Solution

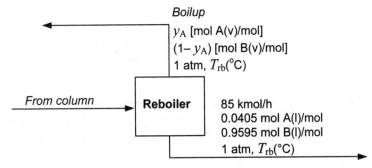

This is just a vapor-liquid equilibrium problem, and does not call for calculation of stream flow rates or for the composition of the feed stream to the reboiler, so the flowchart labeling shown above is sufficient. As before, we use Raoult's law (Eq. 6.4-1 of the text) to relate the liquid-phase and vapor-phase mole fractions of pentane and hexane, and the Antoine equation (Table B.4) for the vapor pressures of the two species.

$$y_A = \frac{x_A p_A^*(T_{rb})}{P} = \frac{0.0405 \times 10^{6.84471 - \frac{1060.793}{T_{rb}+231.541}}}{760}$$

$$y_B = \frac{x_B p_B^*(T_{rb})}{P} = \underline{\hspace{4cm}}$$

(6.60-13)

Eqs. (6.60-13) represent two equations in three unknowns—y_A, y_B, and T_{rb}. A third equation is obtained by noting that the two mole fractions must add up to 1.

$$y_A + y_B = \frac{0.0405 \times 10^{6.84471 - \frac{1060.793}{T_{rb}+231.541}}}{760} + \underline{\hspace{3cm}} = 1$$

(6.60-14)

As was the case with Eq. (6.60-6), this equation can be solved by trial-and-error with a calculator, using a spreadsheet, or with E-Z Solve. If you saved the program from Part (b), you can easily add lines to calculate the expressions for y_A and y_B using copy-and-paste for the Antoine equation formulas, then add them and set the sum equal to 1, and solve for the three variables. Complete the program below for this part of the problem.

(6.60-15)

```
//E-Z Solve Program for Problem 6-60
//Part (b) (fill in using your result from (6.60-7))




//Part (c)
ya = 0.0405*10^(6.84471-1060.793/(Trb+231.541))/760
yb = _____
ya + yb = _____
```

Solution

$$y_A = \underline{\hspace{1.5cm}} \frac{\text{mol A(v)}}{\text{mol}}, \quad y_B = \underline{\hspace{1.5cm}} \frac{\text{mol B(v)}}{\text{mol}}, \quad T_{rb} = \underline{66.6^\circ C}$$

(6.60-16)

(6.60-17)

Q: Examine Eq. (6.4-4) in the text for the bubble point temperature of a liquid mixture with known composition at a pressure P, and compare it with Eq. (6.60-14) for the reboiler temperature. What does the comparison tell you? Why does this result make sense?

A: _____

_____

_____

(d) Calculate the minimum diameter of the pipe connecting the column and the condenser if the maximum allowable velocity in the pipe is 10 m/s. Then list all the assumptions underlying the calculation.

Solution

The volumetric flow rate of a fluid in a pipe, $\dot{V}(m^3/s)$, is the product of the average velocity (m/s) and the cross-sectional area of the flow channel (m^2). For a given value of $\dot{V}$, the smaller the cross section, the greater the velocity. It follows that

$$\dot{V}\left(\frac{m^3}{s}\right) = u_{max}\left(\frac{m}{s}\right) \times \frac{\pi D^2_{min}}{4}(m^2)$$

$$\Rightarrow D^2_{min} = \frac{4\dot{V}}{\pi \cdot u_{max}} = \frac{4}{3.1416}\left|\,\underline{}\,\frac{m^3}{h}\,\right|\,\frac{1\,h}{3600\,s}\,\left|\,\frac{s}{\underline{}\,m}\,\right| = \underline{}\;m^2 \qquad\qquad \textbf{(6.60-18)}$$

$$\Rightarrow D_{min} = \underline{}\;m\,(\underline{}\;cm)$$

$$\qquad\qquad\qquad\qquad\qquad\qquad\qquad\qquad\qquad\qquad\qquad\qquad\qquad\qquad \textbf{(6.60-19)}$$

Q: What assumptions were made in the last calculation?
A: _____

Notes and Calculations

Chapters 4–6 introduced applications of the law of conservation of mass to chemical process systems. As we learned in those chapters, understanding a process requires us to account for material flows into and out of the overall process and the individual units that comprise the process. Chapter 4 defined the basic procedures used for this accounting, and Chapters 5 and 6 showed how different physical properties of process materials and physical and chemical principles governing their behavior could be incorporated into the accounting.

Chapters 7–9 do the same thing for applications of the law of conservation of energy. To design a process and to make it work efficiently and economically, we must also account for the energy that flows into or out of the overall process and the individual process units. Chapter 7 introduces the first law of thermodynamics (which is essentially a mathematical statement of energy conservation) and outlines the basic procedure for writing an energy balance on a process. Chapters 8 and 9 then show how to apply that procedure to (respectively) nonreactive and reactive processes of increasing complexity.

As in Chapters 4–6, we only consider two types of systems in Chapters 7–9:
(a) a closed (batch) process system, in which no material enters or leaves the system until the process is stopped (energy can flow into or out of the system, however)
(b) an open (continuous) process system at steady state

The starting point for solving all energy balance problems in the text (once necessary material balance calculations have been taken care of) is the first law of thermodynamics in either its closed-system or open-system form:

Closed system: $Q - W = \Delta U + \Delta E_k + \Delta E_p$ (terms have units of energy, e.g. J)

Open system: $\dot{Q} - \dot{W}_s = \Delta \dot{H} + \Delta \dot{E}_k + \Delta \dot{E}_p$ (terms have units of power, e.g. J/s)

In these equations, Q(kJ) denotes total heat transferred into the closed system from its surroundings and W is the work done by the system on its surroundings, and $\dot{Q}$(kJ/s) is the rate at which heat is transferred into the open system and $\dot{W}_s$ is the rate at which shaft work (total work minus flow work) is transferred from the system to the surroundings. The derivation of the open system equation from the closed system form is given in Section 7.4c of the text.

For a chemical process with several species entering in the feed streams and leaving in the outlet streams, the terms on the right-hand sides of the two first-law equations are calculated as

$$\Delta U = \sum_{\text{out}} n_i \hat{U}_i - \sum_{\text{in}} n_i \hat{U}_i$$

$$\Delta \dot{H} = \sum_{\text{out}} \dot{n}_i \hat{H}_i - \sum_{\text{in}} \dot{n}_i \hat{H}_i$$

where $\hat{U}_i$(kJ/mol) is the specific internal energy of Species i at its feed or outlet state (temperature, pressure, and phase) and $\hat{H}_i$(kJ/mol) $[= \hat{U}_i + P_i \hat{V}_i]$ is the specific enthalpy at the same state. In this chapter, $\hat{U}_i$ and $\hat{H}_i$ will be given to you directly. In Chapters 8 and 9, you will learn how to calculate them from tabulated data for the species in question.

The procedure for energy balance problems will be to write the appropriate form of the first law, eliminate any terms that are zero or may be neglected, substitute any known values of the remaining variables, and solve the resulting equation for whichever variable is still unknown [which is usually but not always Q (closed system) or $\dot{Q}$ (open system) in the chapter-end problems]. The rules for dropping terms from the first law are given in **Table 7.1**.

Table 7.1: Rules for dropping terms in the first law of thermodynamics

Drop	When	Justification
$Q(\dot{Q})$	System is adiabatic (perfectly insulated) or nearly so or the system and its surroundings are at almost identical temperatures	Heat flows through a conductive medium (as opposed to a thermal insulator) from a higher temperature to a lower one
W or $\dot{W}_s$	System has no moving boundaries or moving parts (pistons, propellers, rotors,...) that can transfer energy to the surroundings, and no other energy transfer mechanisms (e.g. electrical currents or radiation that crosses system boundaries)	
ΔU or $\Delta \dot{H}$	No chemical reactions or phase changes, and process is isothermal or nearly so	Internal energies and enthalpies of species depend strongly on chemical structure, phase, and temperature, and weakly on pressure unless very large pressure changes occur in the process
ΔE_k ($\Delta \dot{E}_k$)	The system is not accelerating (closed) Negligible velocity differences between feed and outlet streams OR there are chemical reactions, phase changes, or temperature changes (open)	In the processes analyzed in this text, as long as $\Delta \dot{H}$ is significant its value will almost invariably be more than an order of magnitude greater than $\Delta \dot{E}_k$
ΔE_p ($\Delta \dot{E}_p$)	The system is not moving up or down (closed). There are no large vertical displacements between the feed and the outlet streams (open)	

We'll remind you of one more thing and then start on problems. As the temperature of a substance increases, the molecules of the substance move faster and therefore have higher energy. Since the internal energy of the substance (U) is the sum of the energies of the individual molecules, a higher temperature means a higher internal energy.

PROBLEM 7.10

A cylinder with a movable piston contains 4.00 liters of a gas at 30°C and 5.00 bar. The piston is slowly moved to compress the gas to 8.00 bar.

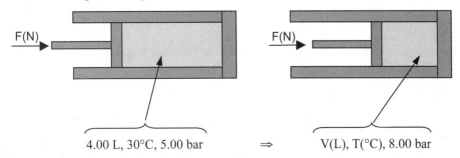

4.00 L, 30°C, 5.00 bar $\Rightarrow$ V(L), T(°C), 8.00 bar

(a) Considering the system to be the gas in the cylinder and neglecting ΔE_p, write and simplify the closed-system energy balance. Do not assume that the gas is isothermal in this part.

Solution

The closed system energy balance equation simplifies as shown below. Fill in the justifications for dropping the two terms shown.

$$Q - W = \Delta U + \Delta E_k + \Delta E_p$$

$$\begin{cases} \Delta E_k = 0 \quad\rule{4cm}{0.4pt} \\ \Delta E_p = 0 \quad\rule{4cm}{0.4pt} \end{cases}$$

(7.10-1)

$$Q - W = \Delta U$$

This is as far as we can go quantitatively without additional information. If we are given values of two of the terms of the equation (as we are in Parts b and c), we can solve for the third one.

(b) Suppose now that the process is carried out isothermally, and the compression work done on the gas equals 7.65 L·bar ($W = -7.65$ L·bar). If the gas is ideal so that $\hat{U}$ is a function only of T, how much heat (in joules) is transferred to or from (state which) the surroundings? (Use the gas constant table in the back of the book to determine the factor needed to convert L·bar to joules.)

Solution

At constant temperature, $\Delta U = 0$. The first law becomes (fill in the blanks):

(7.10-2)

transferred <u>from</u>
gas <u>to</u> surroundings

$$Q - W = \Delta U \xrightarrow[\Delta U=0]{W=-7.65\text{L·bar}} Q = \frac{-7.65 \text{ L} \cdot \text{bar}}{} \left| \frac{\rule{1.5cm}{0.4pt} \text{ J}}{\rule{1.5cm}{0.4pt} \text{ L} \cdot \text{bar}} = -765 \text{ J} \right.$$

In other words, if you compress an ideal gas (transferring energy to it), to keep its temperature constant you must transfer the same amount of energy out of it as heat (that is, cool it).

(c) Suppose instead that the process is adiabatic. Is the final temperature greater than, equal to, or less than 30°C? Explain your reasoning.

Solution

Here is the way to think about it.

- When you push the piston to the right against the restraining force of the gas in the cylinder, you are transferring energy to the gas as work.

- The transferred energy must go someplace, since energy is conserved. The only things that can happen to it are (a) being converted to kinetic or potential energy, (b) being transferred back out of the system as heat, or (c) being converted to internal energy.

- We ruled out (a) in the beginning, and if the process is adiabatic we know that no heat can be transferred, so that the work that went into the gas must go to raise the internal energy of the gas. We have seen that U increases with increasing temperature. If U is higher than it was initially, then T must also be higher, and so the final temperature T must be <u>greater than 30°C</u>.

In short, *compressing a gas adiabatically raises its temperature*. If we knew exactly how $\hat{U}$ depended on T (which we will later in the text), we could calculate ΔU from the energy balance equation and then calculate how much T rises, but since we don't yet know the dependence we have done all we can.

Exercise

Explain why the converse of the behavior just described is also true: If a gas expands against a restraining force, either the gas gets colder or heat must be added to the gas to keep its temperature constant.

(7.10-3)

PROBLEM 7.21

The specific enthalpy of liquid n-hexane at 1 atm varies linearly with temperature and equals 25.8 kJ/kg at 30°C and 129.8 kJ/kg at 50°C.

(a) Determine the equation that relates $\hat{H}$ (kJ/kg) to T(°C) and calculate the reference temperature on which the given enthalpies are based. Then derive an equation for $\hat{U}$ (T)(kJ/kg) at 1 atm.

Solution

The task is to determine a and b in the equation

$$\hat{H} = aT + b \quad : \quad (T_1 = 30,\ \hat{H}_1 = 25.8),\ (T_2 = 50,\ \hat{H}_2 = 129.8)$$

You can substitute the given values and solve for a and b algebraically as shown in Chapter 2, or you can use E-Z Solve to solve the following equations for a and b:

$$25.8 = a*30 + b$$
$$129.8 = a*50 + b$$

Try it, and prove that

$$\underline{\hat{H}\ (\text{kJ/kg}) = 5.20\ T(°C) - 130.2}$$

By definition, at the reference temperature $\hat{H} = 0$. Thus, from the formula just derived,

$$T_{ref} = \underline{} \tag{7.21-1}$$

From Eq. (7.4-7) on p. 321 of the text,

$$\hat{U} = \hat{H} - P\hat{V}$$

Remember that the specific volume $(\hat{V})$ (volume/mass) of a substance is the inverse of the density (mass/volume). From Table B.1 in the text, the specific gravity of liquid hexane is 0.659. Fill in the blanks, using the specific gravity and conversion factors from the inside front cover of the text.

$$\hat{U}\left(\frac{\text{kJ}}{\text{kg}}\right) = \hat{H} - P\hat{V}$$

$$= (5.2T - 130.2)\left(\frac{\text{kJ}}{\text{kg}}\right) - \frac{1\ \text{atm}}{1\ \text{atm}}\left|\frac{\underline{}\ \text{N/m}^2}{}\right|\frac{\text{m}^3}{\underline{}\ \text{kg}}\left|\frac{\underline{}\ \text{J}}{\text{N}\cdot\text{m}}\right|\frac{1\ \text{kJ}}{10^3\ \text{J}} \tag{7.21-2}$$

$$\Rightarrow \quad \underline{\hat{U}(\text{kJ/kg}) = 5.2T(°C) - 130.4}$$

(b) Calculate the average heat transfer rate required to cool 20.0 kg of liquid *n*-hexane from 80°C to 20°C in 5 min.

Solution

Since we are cooling a fixed mass of hexane and not a continuously flowing stream, this is a closed system. The closed system energy balance equation simplifies as shown below:

$$Q - W = \Delta U + \Delta E_k + \Delta E_p$$

$$\begin{cases} W = 0 \quad \text{(Why?)} \; (\underline{\hspace{3cm}}) \\ \Delta E_k = 0 \; (\underline{\hspace{3.5cm}}) \\ \Delta E_p = 0 \; (\underline{\hspace{3.5cm}}) \end{cases} \qquad \text{(7.21-3)}$$

$$Q(\text{kJ}) = \Delta U(\text{kJ}) = m(\text{kg})\left[\hat{U}_{final} - \hat{U}_{initial}\right]\left(\frac{\text{kJ}}{\text{kg}}\right)$$

We can substitute the given value of m and the values of $\hat{U}$ at the initial and final temperatures determined from the equation of Part (a) to calculate the total amount of heat transferred, Q, and then divide Q by the cooling time to determine the average heat transfer rate.

$$\text{(7.21-4)}$$

$$Q(\text{kJ}) = (20.0 \text{ kg})\left[(5.2 \times \underline{\hspace{0.8cm}} \text{-}130.4)\text{-}(5.2 \times \underline{\hspace{0.8cm}} \text{-}130.4)\right]\left(\frac{\text{kJ}}{\text{kg}}\right) = -6240 \text{ kJ}$$

$\Rightarrow$ Average rate of heat removal (kW) = _____

PROBLEM 7.28

Saturated steam at a gauge pressure of 2.0 bar is to be used to heat a stream of ethane. The ethane enters a heat exchanger at 16°C and 1.5 bar (gauge) at a rate of 795 m³/min and is heated at constant pressure to 93°C. The steam condenses and leaves the exchanger as a liquid at 27°C. The specific enthalpy of ethane at the given pressure is 941 kJ/kg at 16°C and 1073 kJ/kg at 93°C.

The heat exchanger may be shown schematically as follows:

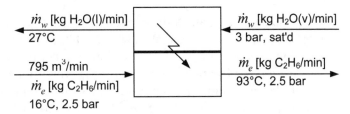

(For simplicity, we have assumed atmospheric pressure is 1 bar when converting the given gauge pressures to absolute pressures.)

The hot steam passes on one side of a metal barrier that is a good conductor of heat, and the cold ethane passes on the other side. Heat is transferred through the barrier from the steam to the ethane at the rate of $\dot{Q}(kW)$. The heat transferred to the ethane raises its temperature to 93°C. The heat transferred from the steam causes the steam to condense and the liquid condensate temperature to drop.

Question

How do we know the steam condenses immediately rather than being cooled to a lower temperature in the vapor phase? Answer below.

(7.28-1)

| |
| |
| |

Following is some information that will help you understand how this process works. (The better you understand it, the less trouble you're likely to have with the calculations.) The problem solution begins on the page after the second set of asterisks.

* * *

- Heat exchange is one of the most common operations in chemical processes, for obvious economic reasons. In the process shown, it is necessary to heat the ethane stream. If we did so by feeding the ethane through a gas-fired, oil-fired, or electrical heater, we would have to pay a high cost for the fuel or power consumption. If elsewhere in the process there is a hot stream that is to be either cooled down or emitted to the atmosphere, a heat exchanger can be used as a preheater for the ethane. Now, instead of having to pay for the energy that heats the cold stream, we recover energy that might otherwise be wasted by transferring it from the hot stream. (The heat exchanger certainly costs something to buy and maintain, but so would the heater.) If we were going to cool the hot stream anyway, the heat exchanger also saves us from having to pay the cost of the cooling.

- We are assuming that the outside of the heat exchanger is perfectly insulated, so that all the heat given up by the steam goes into the ethane. In a real heat exchanger, there would always be some heat transferred through the outer wall. If the insulation is good, however, this leakage would be negligible compared to the heat transfer through the barrier, and so the values we are about to calculate would be quite accurate.

- Our flowchart shows *countercurrent flow*, meaning that the hot and cold streams flow in opposite directions through the unit. *Cocurrent flow*, in which the streams flow in the same direction, would not work for this process. Can you figure out why? (We'll supply the answer later in the problem.)

- In a real heat exchanger the steam and ethane streams would not simply flow on opposite sides of a wall. For example, one of the fluids might flow through the insides of a bundle of many narrow tubes mounted in a large cylindrical shell while the other might flow through this shell (on the outside of the tubes). The rate of heat transfer is proportional to the surface area of the interface between the two fluids. The large surface/volume ratio of a collection of small tubes means that a much smaller (in volume) and less expensive unit is required to get the same rate of heat transfer. For a picture of what this would look like, see the figure attached to workbook Problem 4.36 or the section on shell-and-tube heat exchangers in the *Visual Encyclopedia of Chemical Engineering Equipment* on the CD that came with the text.

- The description above applies to a "shell-and-tube" heat exchanger. There are many different configurations of shell-and-tube heat exchangers—they vary primarily in differences in the complexity of the flow on the shell side of the heat exchanger. There are other types of heat exchangers as well. For example, look up the "plate-and-frame" type exchanger on the internet and compare its contacting configuration to that of the shell-and-tube exchanger. Each type of heat exchanger has its own design nuances with higher levels of design complexity accompanying more complex flow contacting patterns.

* * *

(a) How much energy (kW) must be transferred to the ethane to heat it from 16°C to 93°C?

Solution

There are two different process subsystems that we could write energy balances on in addition to the complete heat exchanger: the ethane heating process and the steam condensation and cooling process. The flowcharts for the subsystems appear as follows:

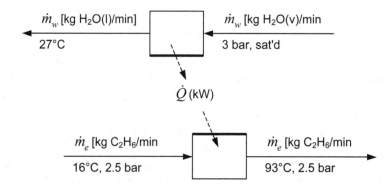

Look at the three flowcharts (the complete heat exchanger, the ethane heating process, and the steam cooling and condensation process), and answer the following questions about the degree-of-freedom analyses.

Ethane Heating Process Questions (fill in the blanks)   **(7.28-2)**

Q: Why can't you count any material balances for any of the processes?

A: _____

Q: Why doesn't $\dot{Q}$ count as an unknown variable for the DOF analysis of the exchanger?

A: _____

Q: How many degrees of freedom does each process system have?

A: **Overall exchanger** ___ variables (_____) – 1 gas eq. of state – 1 energy balance = 0 DF

   Ethane heating ___ variables (_____) – _____ = 0 DF

   Steam condensation/cooling ___ variables (_____) – _____ = 1 DF

Q: What strategy should be followed to calculate all unknown variables?

A: (i) _____

_____

(ii) _____

_____

The open system energy balance on the ethane heating process is written below. Using information on the flowchart, fill in the correct values for the gas variables (and conversion factors).

$$\dot{m}_e \left(\frac{\text{kg C}_2\text{H}_6}{\text{min}} \right) = \dot{n}_e \left(\frac{\text{kmol C}_2\text{H}_6}{\text{min}} \right) \times 30.07 \left(\frac{\text{kg C}_2\text{H}_6}{\text{kmol}} \right) = \left(\frac{P\dot{V}}{RT} \right)(30.07)$$

$$= \frac{\underline{\quad} \text{ bar}}{\underline{\quad} \text{ K}} \left| \frac{\underline{\quad} \text{ m}^3}{\text{min}} \right| \frac{\underline{\quad} \text{ L}}{\text{m}^3} \left| \frac{1}{\underline{\quad}} \frac{\text{mol} \cdot \text{K}}{\text{L} \cdot \text{bar}} \right| \frac{\underline{\quad} \text{ g}}{\text{mol}} \left| \frac{1 \text{ kg}}{10^3 \text{ g}} \right.$$ **(7.28-3)**

$$= 2.487 \times 10^3 \text{ kg C}_2\text{H}_6/\text{min}$$

$$\dot{Q} - \dot{W}_s = \Delta\dot{H} + \Delta\dot{E}_k + \Delta\dot{E}_p$$

$$\left\{ \begin{array}{l} W_s = 0 \quad (\text{Why?}) \, (\underline{\hspace{3cm}}) \\ \Delta E_k = 0 \, (\underline{\hspace{3cm}}) \\ \Delta E_p = 0 \, (\underline{\hspace{3cm}}) \end{array} \right.$$ **(7.28-4)**

$$\dot{Q} = \Delta\dot{H} = \dot{m}_e \left[\hat{H}_{out} - \hat{H}_{in} \right]$$

$$= \frac{\underline{\quad} \text{ kg}}{\text{min}} \left| \frac{(\underline{\quad} - \underline{\quad}) \text{ kJ}}{\text{kg}} \right| \frac{1 \text{ min}}{60 \text{ s}} \left| \frac{1 \text{ kW}}{1 \text{ kJ/s}} \right. = \underline{\hspace{2cm}}$$

Name: _____

Date: _____

(b) Assuming that all the energy transferred from the steam goes to heat the ethane, at what rate in m^3/s must steam be supplied to the exchanger? If the assumption is incorrect, would the calculated value be too high or too low?

Strategy

An energy balance on the steam condensation/cooling process is required. We know the flow rate of the ethane and its temperature change so we can calculate the amount of energy needed to heat this gas. All the energy is assumed to come from the steam. From the conditions of the steam, we can determine how much energy per kg of steam is released when the steam condenses and cools and, therefore, how much steam is required to deliver the computed amount of energy to the ethane.

In the energy balance equation, we will need values for the specific enthalpy of steam at the inlet condition (saturated, 2.0 bar) and of liquid water at the outlet condition (liquid, 27°C), and to convert the mass flow rate of the inlet steam to the corresponding volumetric flow rate we'll need the specific volume of the steam.

(7.28-5)

Q: Where in the book can we find the required values of $\hat{H}$ and $\hat{V}$ for the inlet steam?

A: _____

Q: Where in the book can we find the required value of $\hat{H}$ for the outlet liquid?

A: _____

Q: The value of $\hat{H}$ in Table B.5 for liquid water at 27°C applies to a pressure that is probably different from the outlet water pressure in the process (which we don't know). Why doesn't it matter?

A: _____

The open system energy balance for this system is (fill in the blanks):

$$\dot{Q} - \dot{W}_s = \Delta\dot{H} + \Delta\dot{E}_k + \Delta\dot{E}_p$$

$$\begin{cases} \dot{W}_s = 0 \quad \text{(no moving parts)} \\ \Delta\dot{E}_k = 0 \ \left(\text{neglect kinetic energy change from inlet to outlet}\right) \\ \Delta\dot{E}_p = 0 \ \left(\text{neglect vertical displacement between inlet and outlet}\right) \end{cases}$$

$$\dot{Q} = \Delta\dot{H} = \dot{m}_w \left[\hat{H}_{out} - \hat{H}_{in} \right]$$

$$\Rightarrow \quad - \underline{\hspace{1cm}} \ \text{kW} = \frac{\dot{m}_w(\text{kg/s}) \left| (\underline{\hspace{0.8cm}} - \underline{\hspace{0.8cm}}) \ \text{kJ} \right| 1 \ \text{kW}}{\left| \text{kg} \right| 1 \ \text{kJ/s}}$$

(7.28-6)

$$\Rightarrow \dot{m}_w = \underline{\hspace{1cm}} \ \text{kg/s} \Rightarrow \dot{V}_{steam} = \dot{m}_w \left(\underline{\hspace{0.5cm}}\frac{\text{kg}}{\text{s}}\right) \hat{V}_{steam} \left(\underline{\hspace{0.5cm}}\frac{m^3}{\text{kg}}\right) = \left(\underline{\hspace{0.5cm}}\frac{\text{kg}}{\text{s}}\right)\left(\underline{\hspace{0.5cm}}\frac{m^3}{\text{kg}}\right)$$

$$= \underline{\hspace{1cm}} \ m^3/s$$

If heat is lost to the surroundings, the calculated value of $\dot{m}_w$ is too _____.   (7.28-7)

Explain: _____

(c) Should the heat exchanger be set up for cocurrent or countercurrent flow? Explain. (*Hint*: Remember that heat always flows from a higher temperature to a lower temperature.)

Solution

The heat exchanger should use <u>countercurrent flow</u>. For cocurrent flow the exchanger would appear as follows:

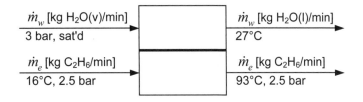

$\dot{m}_w$ [kg H$_2$O(v)/min]
3 bar, sat'd

$\dot{m}_e$ [kg C$_2$H$_6$/min]
16°C, 2.5 bar

$\dot{m}_w$ [kg H$_2$O(l)/min]
27°C

$\dot{m}_e$ [kg C$_2$H$_6$/min]
93°C, 2.5 bar

(7.28-8)

The problem with this setup is _____

Notes and Calculations

PROBLEM 7.41

Air at 38°C and 97% relative humidity is to be cooled to 18°C and fed into a plant area at a rate of 510 m³/min.

(a) Draw and label the flowchart, do the degree-of-freedom analysis, and write in an efficient order the equations you would use to calculate all of the unknown variables on the chart. For now, assume that you will be able to determine the specific enthalpies of water and air at the inlet and outlet conditions.

(b) Calculate the rate (kg/min) at which water condenses.

(c) Calculate the cooling requirement in tons (1 ton of cooling = 12,000 Btu/h), assuming that the enthalpy of water vapor is that of saturated steam at the same temperature and the enthalpy of dry air is given by the expression

$$\hat{H}(\text{kJ/mol}) = 0.0291[T(°C) - 25]$$

(Justify the assumption about the enthalpy of the water vapor.)

Solution

(a) Assume all streams are at 1 atm. Whenever some of the water vapor in a humid air stream is condensed in a process and the air stream leaves the process in thermal equilibrium with the condensate, the exiting humid air may be assumed to be saturated at the stream temperature. We've labeled the humid air stream leaving the process accordingly.

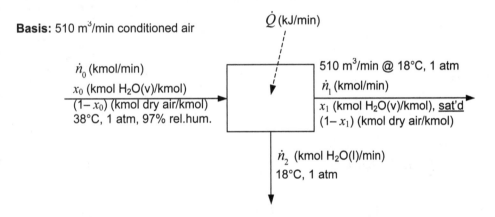

Basis: 510 m³/min conditioned air

$\dot{Q}$ (kJ/min)

$\dot{n}_0$ (kmol/min)
x_0 (kmol H₂O(v)/kmol)
(1– x_0) (kmol dry air/kmol)
38°C, 1 atm, 97% rel.hum.

510 m³/min @ 18°C, 1 atm
$\dot{n}_1$ (kmol/min)
x_1 (kmol H₂O(v)/kmol), sat'd
(1– x_1) (kmol dry air/kmol)

$\dot{n}_2$ (kmol H₂O(l)/min)
18°C, 1 atm

(7.41-1)

<table>
<tr><td colspan="2" align="center">DEGREE-OF-FREEDOM ANALYSIS</td></tr>
<tr><td align="center">UNKNOWNS AND INFORMATION</td><td align="center">JUSTIFICATION/CONCLUSION</td></tr>
<tr><td>+ 6 unknowns</td><td>$\dot{n}_0, x_0, \dot{n}_1, x_1, \dot{n}_2, \dot{Q}$</td></tr>
<tr><td>− 2 material balances</td><td>_____</td></tr>
<tr><td>− 1 _____</td><td>Solve for x_0</td></tr>
<tr><td>− 1 _____</td><td>Solve for $\dot{n}_1$</td></tr>
<tr><td>− 1 _____</td><td>Solve for x_1</td></tr>
<tr><td>− 1 energy balance</td><td>Solve for $\dot{Q}$</td></tr>
<tr><td> 0 degrees of freedom</td><td align="center">Problem is solvable</td></tr>
</table>

Strategy

Write the equations in the order that minimizes the number that have to be solved simultaneously, and indicate where you would find needed physical properties (vapor pressures, specific enthalpies,...). In the equations that follow, the subscript W denotes water—for example, $p_W^*(T)$ denotes the vapor pressure of water at temperature T.

(7.41-2)

Relative humidity of gas at inlet $\boxed{x_0}P_0 = 0.97\,p_W^*(38°C)$ [7.41-1]

1 atm   Table B.5

Raoult's law at outlet   [7.41-2]

Ideal gas EOS   [7.41-3]

_____ **balance**   [7.41-4]

_____ **balance**   [7.41-5]

Energy balance   [7.41-6]

$$\dot{Q} = \Delta\dot{H} = \sum_{out} \dot{n}_i \hat{H}_i - \sum_{in} \dot{n}_i \hat{H}_i \quad (\text{Neglect} \underline{\hspace{4cm}})$$

$$\boxed{\dot{Q}} = \dot{n}_1 x_1 \hat{H}_{W(v,18°C)} + \dot{n}_1(1-x_1)\hat{H}_{DA(18°C)} + \dot{n}_2 \hat{H}_{W(l,18°C)} - \dot{n}_0 x_0 \hat{H}_{W(v,38°C)} - \dot{n}_0(1-x_0)\hat{H}_{DA(38°C)}$$

$\hat{H}_W$ from Table _____, $\hat{H}_{DA}$ from given formula

Once we have solved the first five equations, we can convert $\dot{n}_2$ to the mass flow rate of liquid water in kg/min to solve Part (b), and we can then solve the energy balance equation and convert $\dot{Q}$ to tons to solve Part (c). Complete the calculations below.

(b) $x_0 = $ _____ $= 0.0634$ kmol H_2O/kmol   **(7.41-3)**

$x_1 = $ _____ $= $ _____ kmol H_2O/kmol   **(7.41-4)**

$$\dot{n}_1 = \frac{P_1 \dot{V}_1}{RT_1} = \frac{\underline{\quad} \text{ atm}}{\underline{\quad} \text{ K}} \bigg| \frac{\underline{\quad} \text{ m}^3}{\text{min}} \bigg| \frac{\text{kmol} \cdot \text{K}}{\underline{\quad} \text{ m}^3 \cdot \text{atm}} = 21.36 \text{ kmol/min} \qquad \textbf{(7.41-5)}$$

Dry air balance [   ] $\Rightarrow \dot{n}_o = 22.34 \text{ kmol/min}$ **(7.41-6)**

Water balance [   ] **(7.41-7)**

$$\Rightarrow \dot{n}_2 = \underline{\quad} \text{ kmol/min} \Rightarrow \dot{m}_2 = \frac{\underline{\quad} \text{ kmol}}{\text{min}} \bigg| \frac{\underline{\quad} \text{ kg}}{1 \quad \text{kmol}}$$

$$= \underline{\quad} \text{ kg/min } H_2O \text{ condenses}$$

(c) First, we evaluate the enthalpies needed. **(7.41-8)**

$$\hat{H}_{air}(38°C) = 0.0291(38 - 25) = 0.378 \text{ kJ/mol}$$

$$\hat{H}_{air}(18°C) = \underline{\hspace{3cm}} = \underline{\hspace{2cm}} \text{ kJ/mol}$$

$$\hat{H}_{H_2O}(v, 38°C) = \frac{\underline{\quad} \text{ kJ}}{\text{kg}} \bigg| \frac{1 \text{ kg}}{10^3 \text{ g}} \bigg| \frac{18.02 \text{ g}}{\text{mol}} = 46.33 \text{ kJ/mol}$$

$$\hat{H}_{H_2O}(v, 18°C) = \frac{\underline{\quad} \text{ kJ}}{\text{kg}} \bigg| \frac{1 \text{ kg}}{10^3 \text{ g}} \bigg| \frac{18.02 \text{ g}}{\text{mol}} = \underline{\quad} \text{ kJ/mol} \Bigg\} \text{Table B.5}$$

$$\hat{H}_{H_2O}(l, 18°C) = \frac{\underline{\quad} \text{ kJ}}{\text{kg}} \bigg| \frac{1 \text{ kg}}{10^3 \text{ g}} \bigg| \frac{18.02 \text{ g}}{\text{mol}} = \underline{\quad} \text{ kJ/mol}$$

Add the missing terms to complete the energy balance. (Eq. [7.41-6]) **(7.41-9)**

$$\dot{Q} = (\dot{n}\hat{H})_{W(v,18°C)} + (\dot{n}\hat{H})_{DA(18°C)} + (\dot{n}\hat{H})_{W(l,18°C)} - (\dot{n}\hat{H})_{W(v,38°C)} - (\dot{n}\hat{H})_{DA(38°C)}$$

$$= \left(21.36 \frac{\text{kmol}}{\text{min}}\right)\left(10^3 \frac{\text{mol}}{\text{kmol}}\right)\left(0.0204 \frac{\text{mol } H_2O}{\text{mol}}\right)\left(45.67 \frac{\text{kJ}}{\text{mol } H_2O}\right) + \left(21.36 \times 10^3\right)\left(1 - 0.0204\right)\left(-0.204\right)$$

$$+$$

$$+$$

$$= \frac{\underline{\quad} \text{ kJ}}{\text{min}} \bigg| \frac{\underline{\quad} \text{ min}}{1 \text{ hr}} \bigg| \frac{\underline{\quad} \text{ Btu}}{1 \text{ kJ}} \bigg| \frac{1 \text{ ton cooling}}{-12,000 \text{ Btu/h}} = \underline{\quad} \text{ tons of cooling}$$

The values of $\hat{H}_{H_2O}$ used in this calculation are for saturated liquid and vapor, which from Table B.5 are at P = 0.0662 bar (38°C) and P = 0.02062 bar (18°C). Those are far from the actual system pressures at the inlet and outlet and they are probably not the correct pressures at which to evaluate $\hat{H}_{H_2O}$, but we used them anyway without worrying about the error we might be introducing. The reason is that at pressures on the order of 10 atm (or 10 bar) and lower, specific internal energies and enthalpies of species are almost independent of pressure. To prove this to yourself, look up (Table B.7) and record the values of $\hat{U}$ and $\hat{H}$ at 100°C and several different pressures in the following table:

(7.41-10)

Pressure	0.1 bar	0.5 bar	1.0 bar	5.0 bar	10.0 bar
Phase	vapor	vapor	vapor	liquid	liquid
$\hat{U}$ (kJ/kg)	2516	_____	_____	_____	_____
$\hat{H}$ (kJ/kg)	2688	_____	_____	_____	_____

Clearly, internal energy and enthalpy depend significantly on phase, and if you look them up at temperatures different from 100°C you will find that they also depend significantly on temperature. However, using values at any of the listed pressures for water at 100°C would introduce an insignificant error if the actual pressure is different, as long as P<10 bar.

PROBLEM 7.45

A wet steam at 20 bar with a quality of 0.97 (see text Problem 7.32) leaks through a defective steam trap and expands to a pressure of 1 atm. The process can be considered to take place in two stages: a rapid adiabatic expansion to 1 atm accompanied by complete evaporation of the liquid droplets in the wet steam, followed by cooling at 1 atm to ambient temperature. $\Delta \dot{E}_k$ may be neglected in both stages. [*Note:* As described in Problem 7.34 in the text, a steam trap is a device that separates liquid water condensate from uncondensed steam. The fact that water vapor is included in the effluent from the steam trap is evidence that the trap is defective.]

(a) Estimate the temperature of the superheated steam immediately following the rapid adiabatic expansion.

Solution

According to Problem 7.32, a wet steam is a mixture of liquid droplets and saturated water vapor, and the quality of a wet steam is the fraction by mass that is vapor. The flowchart for this process may be drawn as follows.

Basis: 1 kg wet steam

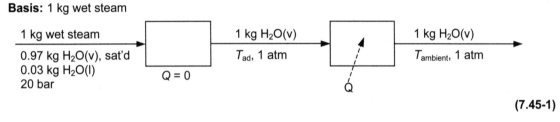

$$(7.45\text{-}1)$$

Q. What is the temperature of the wet steam (the feed to the adiabatic expansion stage)?
A. $T_{feed} =$ _____ (from Table _____)

In the energy balance problems we have done so far, inlet and outlet temperatures were known and the problem was to solve the energy balance equation for Q. In problems on adiabatic processes such as this one, Q is given (it is zero by definition) and (usually) the outlet condition is to be determined. The energy balance is written in the same way as before, except that the specific enthalpies ($\hat{H}$) of all outlet species are left as unknowns or expressed as functions of the outlet temperature.

Before we do any calculations, let's think a bit about the adiabatic expansion. Remember that two processes are occurring simultaneously: evaporation of the liquid in the wet steam, and expansion of the steam from 20 bar (close to 20 atm) to 1 atm.

$$(7.45\text{-}2)$$

Q. What would you anticipate regarding the temperature following the adiabatic expansion?
____ $T_{ad} < T_{feed}$ ____ $T_{ad} = T_{feed}$ ____ $T_{ad} > T_{feed}$ ____ No way to know
A. Explain your answer. _____

Now to the calculation. Knowing that we will need to substitute the specific enthalpies of the feed and outlet stream components in the energy balance, let's find them now.

Specific Enthalpies

Inlet

$$\left.\begin{array}{l} \text{Vapor: } \hat{H}_v = \underline{\hspace{1cm}} \text{ kJ/kg} \\ \text{Liquid: } \hat{H}_l = \underline{\hspace{1cm}} \text{ kJ/kg} \end{array}\right\} (\text{Table} \underline{\hspace{0.6cm}})$$

(7.45-3)

Outlet $\hat{H}(T_{ad}, 1 \text{ atm}) = \hat{H}_{ad}$ (Since we don't know T, we can't look up $\hat{H}$.)

Energy balance on the adiabatic expansion stage

$$\dot{Q} - \dot{W}_s = \Delta\dot{H} + \Delta\dot{E}_k + \Delta\dot{E}_p$$

$$\begin{cases} \dot{Q} = 0 \text{ (adiabatic)}, \ \dot{W}_s = 0 \text{ (no moving parts)}, \ \Delta\dot{E}_k = 0 \ \left(\text{given as negligible}\right) \\ \Delta\dot{E}_p = 0 \ \left(\text{neglect vertical displacement between inlet and outlet}\right) \end{cases}$$

$$0 = \Delta\dot{H} = \sum_{\text{out}} \dot{n}_i \hat{H}_i - \sum_{\text{in}} \dot{n}_i \hat{H}_i$$

(7.45-4)

$$\Rightarrow 0 = (1 \text{ kg})\left[\hat{H}_{ad}\left(\frac{\text{kJ}}{\text{kg}}\right)\right] - (\underline{\hspace{0.5cm}} \text{ kg})\left(\underline{\hspace{1cm}} \frac{\text{kJ}}{\text{kg}}\right) - (\underline{\hspace{0.5cm}} \text{ kg})\left(\underline{\hspace{1cm}} \frac{\text{kJ}}{\text{kg}}\right)$$

$$\Rightarrow \hat{H}_{ad} = \underline{\hspace{1cm}} \frac{\text{kJ}}{\text{kg}}$$

We now know that at the outlet of the first stage, $P = 1$ atm (≈ 1 atm) and $\hat{H} =$ the value just calculated. Interpolating in Table B.7 in the text, we find $T_{ad} \approx$ _____ °C   (7.45-5)

(b) Someone looking at the steam trap would see a clear space just outside the leak and a white plume forming a short distance away. (The same phenomenon can be observed outside the spout of a kettle in which water is boiling.) Explain this observation. What would the temperature be at the point where the plume begins?

Solution

(7.45-6)

T = _____.

Here again is the first law of thermodynamics (also known as the energy balance equation) in its closed-system and open-system forms:

Closed system $Q - W = \Delta U + \Delta E_k + \Delta E_p$ [8.1]

Open system $\dot{Q} - \dot{W}_s = \Delta \dot{H} + \Delta \dot{E}_k + \Delta \dot{E}_p$ [8.2]

where

$$\Delta U = \sum_{out} n_i \hat{U}_i - \sum_{in} n_i \hat{U}_i \qquad [8.3]$$

$$\Delta \dot{H} = \sum_{out} \dot{n}_i \hat{H}_i - \sum_{in} \dot{n}_i \hat{H}_i \qquad [8.4]$$

In Chapter 7, when we needed a specific internal energy ($\hat{U}$) or a specific enthalpy ($\hat{H}$) of a process material at its inlet or outlet condition to substitute into Eq. [8.3] or [8.4], respectively, we had two ways to find it: (1) from a formula given in the problem statement, or (2) if the species was liquid water or water vapor, by looking up $\hat{U}$ or $\hat{H}$ in the steam tables (Tables B.5–B.7). We next learn how to determine $\hat{U}$ and $\hat{H}$ for any species at any condition. Knowing that, we will be able to write and solve the closed- or open-system energy balance equation for any process. In this chapter, we consider only processes that do not involve chemical reactions, and in Chapter 9 we will work problems that deal with reactive processes.

The absolute value of $\hat{U}$ for a species at a given condition (which is the total energy of the translation, rotation, vibration, etc., of individual molecules of the species) can never be known (and neither can $\hat{H} = \hat{U} + P\hat{V}$); all we can determine is the difference between values at different conditions. Therefore, when we determine values of $\hat{U}$ or $\hat{H}$ for a species to substitute into Eq. [8.3] or [8.4], they must always be *relative* to that species at a specified reference condition. The value of, say, $\hat{H}$ that we calculate for Species A at its inlet or outlet condition (temperature, pressure, and phase) would actually be $\Delta \hat{H}$ for the process A (reference condition) → A (process condition). The choice of reference condition is strictly a matter of convenience: the calculated value of ΔU or $\Delta \dot{H}$ will be the same regardless of the reference conditions chosen for each process species.

To calculate $\Delta \hat{U}$ or $\Delta \hat{H}$ for a process in which a species at a reference condition goes to another specified condition [e.g., for the process $H_2O(l, 25°C, 1 \text{ atm})$ → $H_2O(v, 300°C, 10 \text{ atm})$], we create a hypothetical *process path* from the initial state to the final state, comprising steps for which we can calculate $\Delta \hat{U}$ or $\Delta \hat{H}$ from tabulated data. For nonreactive species, which is all we consider in this chapter, there are four types of steps we can use:

(1) Change the temperature of a component(s) from T_1 to T_2 at constant pressure and phase

$$\Delta \hat{H} = \int_{T_1}^{T_2} C_p dT \quad \text{[Look up } C_p \text{ in Table B.2]} \qquad [8.5]$$

$$\Delta \hat{U} = \int_{T_1}^{T_2} C_v dT \qquad [8.6]$$

where $C_v = C_p - R$ (ideal gases)

$C_v \approx C_p$ (liquids and solids)

A fast alternative to integrating C_p manually in Eq. [8.5] is to determine $\Delta \hat{H}$ directly from the physical property tables in the CD that came with the text.

(2) Change the pressure of a component(s) from P_1 to P_2 at constant temperature and phase

$\Delta \hat{U} \approx 0$ [solids, liquids, and gases at close to ideal gas conditions]   [8.7]

$\Delta \hat{H} \approx 0$ [gases at close to ideal gas conditions]   [8.8]

$\quad \approx \hat{V} \Delta P$ [solids and liquids]   [8.9]

(3) Undergo a phase change at constant temperature and pressure

Melting: $\Delta \hat{H} = \Delta \hat{H}_m (T, P)$ [*Heat of melting* or *heat of fusion*]   [8.10]

$\quad\quad\quad \Delta \hat{U} = \Delta \hat{U}_m (T, P) \approx \Delta \hat{H}_m (T, P)$ [8.11]

Look up $\Delta \hat{H}_m$ at 1 atm and the melting point of the species at 1 atm in Table B.1.

Vaporization: $\Delta \hat{H} = \Delta \hat{H}_v (T, P)$ [*Heat of vaporization*]   [8.12]

$\quad\quad\quad\quad \Delta \hat{U} \approx \Delta \hat{H}_v (T, P) - RT$ [Vapor at nearly ideal gas conditions)   [8.13]

Look up $\Delta \hat{H}_v$ at 1 atm and the boiling point of the species at 1 atm in Table B.1.

(4) Combine (mix) a solute and solvent to form a solution at constant temperature and pressure

Heats of mixing and heats of solution for HCl, H_2SO_4, and NaOH at 25°C and 1 atm are listed in Table B.11 in the text; data for other solutes at given dilutions may be found in Perry's Handbook.

PROBLEM 8.9

Chlorine gas is to be heated from 100°C and 1 atm to 200°C.

(a) Calculate the heat input (kW) required to heat a stream of the gas flowing at 5.0 kmol/s at constant pressure.

Solution

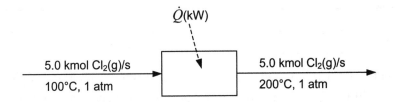

The energy balance for this open process neglecting shaft work and kinetic and potential energy changes is

$$\dot{Q} = \Delta\dot{H} = \sum_{\text{out}} \dot{n}_i \hat{H}_i - \sum_{\text{in}} \dot{n}_i \hat{H}_i$$

$$= \dot{n}_{Cl_2} \Delta\hat{H} = \dot{n}_{Cl_2} [\hat{H}(Cl_2, 200°C, 1 \text{ atm}) - \hat{H}(Cl_2, 100°C, 1 \text{ atm})]$$

For the process of heating a gas at constant temperature, the formula for $\Delta\hat{H}$ is given by Eq. [8.5] on p. 8-1:

(8.9-1)

$$\Delta\hat{H}\left(\frac{kJ}{mol}\right) = \int \underline{\quad} \left(\underline{\quad}\right)_{Cl_2} dT$$

$$\xrightarrow{\text{Table B.2}} \Delta\hat{H}\left(\frac{kJ}{mol}\right) = \int_{100}^{200} \left[0.03360 + 1.367 \times 10^{-5} T - \underline{\quad\quad} T^2 + \underline{\quad\quad} T^3\right] dT$$

You can do this calculation in either of two ways:

(1) Integrate the polynomial, substitute the upper and lower temperature limits into the resulting expression, and subtract the second value from the first one. Or, if you don't have a lot of spare time,

(2) open *Interactive Chemical Process Principles,* bring up the Physical Property Database, click on the "Enthalpies" tab, select chlorine as the species, 100 and 200 as Temp1 and Temp2, respectively, and °C as the temperature unit. Then, with the cursor right after the 200, hit the Enter key and read the value of $\Delta\hat{H}$.

Solution: $\Delta\hat{H} = $ _____kJ/mol   (8.9-2)

It follows that

$$\dot{Q} = \dot{n}_{Cl_2} \Delta\hat{H} = \frac{\underline{\quad} \text{ mol}}{s} \left| \frac{\underline{\quad} \text{ kJ}}{\text{mol}} \right| \frac{1 \text{ kW}}{1 \text{ kJ/s}} = \underline{\quad} \text{ kW}$$ (8.9-3)

Name: _____

Date: _____

(8.9-4)

> **Q:** Suppose we had not told you that the process was taking place at constant pressure, but instead had said that the pressure at the outlet was lower than 1 atm. Would we have to change the calculation? If so, how? Explain.
>
> **A:** _____
>
> _____
>
> _____

(b) Calculate the heat input (kJ) required to raise the temperature of 5.0 kmol of chlorine in a closed rigid vessel from 100°C and 1 atm to 200°C. (*Suggestion*: Evaluate $\Delta \hat{U}$ directly from the result of the first calculation, so that you do not have to perform another integration.) What is the physical significance of the numerical difference between the values calculated in Parts (a) and (b)?

Solution:

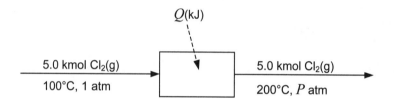

Q(kJ)

5.0 kmol Cl$_2$(g)
100°C, 1 atm

5.0 kmol Cl$_2$(g)
200°C, P atm

(8.9-5)

> **Q:** What can you say about the final pressure, P(atm)? Explain your answer.
>
> ___$P<1$ ___$P=1$ ___$P>1$ ___Can't tell.
>
> **A:** _____

For this closed system with no work or changes in kinetic or potential energy, the energy balance is

$$Q = \underline{\hspace{1cm}} = n_{Cl_2} \Delta \hat{U} \qquad (8.9\text{-}6)$$

The process taking place is almost the same as that of Part (a)—raising the temperature of chlorine gas from 100°C to 200°C. The final pressures in the two processes are different, but the effect of the pressure change on the values of $\Delta \hat{U}$ and $\Delta \hat{H}$ at the low pressures involved should be negligible, and so we can assume that the value of $\Delta \hat{H}$ calculated in Part (a) of 3.53 kJ/mol also applies here. Our goal is then to express $\Delta \hat{U}$ in terms of $\Delta \hat{H}$. That is not a hard task, fortunately, remembering the definition of $\hat{H}$ as $\hat{U} + P\hat{V}$. We can write

$$\Delta \hat{U} = \Delta(\hat{H} - P\hat{V}) = (\hat{H}_2 - P_2\hat{V}_2) - (\hat{H}_1 - P_1\hat{V}_1) = (\hat{H}_2 - \hat{H}_1) - (P_2\hat{V}_2 - P_1\hat{V}_1)$$

Make the reasonable assumption that the chlorine behaves like an ideal gas and substitute for the $P\hat{V}$ terms in the last term on the right to fill in the missing terms in the rest of the calculation in **(8.9-7)** below.

$$\Delta \hat{U} = \Delta \hat{H} - (P_2 \hat{V}_2 - P_1 \hat{V}_1) \xrightarrow{P\hat{V} = RT} \Delta \hat{U} = \Delta \hat{H} - \underline{\qquad}$$

$$\Rightarrow Q = \Delta U = n_{Cl_2} \Delta \hat{U} = (\underline{\qquad} \text{ mol}) \left[\underline{\qquad} \frac{kJ}{mol} - \frac{\underline{\qquad} J}{mol \cdot K} \left| (\underline{\quad} - \underline{\quad}) K \right| \frac{1 \text{ kJ}}{10^3 \text{ J}} \right] \qquad \textbf{(8.9-7)}$$

$$= \underline{\qquad} kJ$$

(8.9-8)

Q: In Part (a), we calculated that raising the temperature of 5 kmol (the amount of chlorine processed in one second) from 100°C to 200°C in a continuous process required 17,650 kJ of heat. We just now calculated that to raise the temperature of 5 kmol of chlorine from 100°C to 200°C in a batch process required only 13,500 kJ. Where does the extra 4,150 kJ go in the continuous process?

A: _____

_____

_____

_____

(c)

(8.9-9)

Q: To accomplish the heating of Part (b), we would actually have to supply an amount of heat to the vessel greater than the amount calculated. Why?

A: _____

_____

_____

_____

Notes and Calculations

PROBLEM 8.18

A liquid mixture of 30.0 wt% acetone and 70.0 wt% 2-methyl-1-pentanol ($C_6H_{14}O$) is cooled from 45°C to 20°C at a rate of 450 kg/min. Calculate the required cooling rate (kW).

Solution

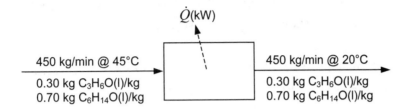

450 kg/min @ 45°C
0.30 kg $C_3H_6O(l)$/kg
0.70 kg $C_6H_{14}O(l)$/kg

$\dot{Q}$(kW)

450 kg/min @ 20°C
0.30 kg $C_3H_6O(l)$/kg
0.70 kg $C_6H_{14}O(l)$/kg

Since no material balance calculations are necessary, we may proceed directly to the energy balance. Whenever more than one species is involved in an open-system process, construct an inlet-outlet enthalpy table to help you keep track of the $\dot{n}$'s and $\hat{H}$'s of the feed and outlet stream components (or an inlet-outlet internal energy table if ΔU is to be calculated as part of a closed system balance). If a species is at its reference state, enter 0 for $\hat{H}$; otherwise label its specific enthalpy in the table as shown below.

Begin by choosing reference states for each process species. When we plan to take specific enthalpies directly from a table (such as the steam tables for water), the most convenient reference is usually the one used to generate the table; otherwise, it is most convenient to choose an inlet or outlet state so that at least one entry can be set equal to zero with no calculations required. For illustrative purposes, we will choose as reference states the inlet state for acetone and the outlet state for the methyl pentanol (normally, we would use either the inlet state or outlet state for both—the choice does not affect the final answer).

References: $C_3H_6O(l,45°C,1atm)$, $C_6H_{14}O(l,20°C,1atm)$

	$\dot{n}_{in}$ (kg/min)	$\hat{H}_{in}$ (kJ/kg)	$\dot{n}_{out}$ (kg/min)	$\hat{H}_{out}$ (kJ/kg)
$C_3H_6O(l)$	135	$\hat{H}_1$	135	$\hat{H}_3$
$C_6H_{14}O(l)$	315	$\hat{H}_2$	315	$\hat{H}_4$

The strategy will be to determine the four specific enthalpies in the table, and then substitute them into the energy balance to calculate $\dot{Q}$.

$$\dot{Q} = \Delta\dot{H} = \sum_{out} \dot{n}_i \hat{H}_i - \sum_{in} \dot{n}_i \hat{H}_i$$

(8.18-1)

Q: Do you expect the sign of $\dot{Q}$ to be positive or negative, or is there no way to tell?

A: _____ . Why? _____

Q: The problem statement says nothing about the pressure of the process streams. Is that a problem? If not, why not?

A: _____

Before doing any calculations, let's make sure we're on the same page regarding the meaning of the entries in the enthalpy table. Consider the reference states defined at the top of the enthalpy table, and fill in the blanks.

(8.18-2)

- $\hat{H}_2$ is the specific enthalpy of $C_6H_{14}O(l,\_\_\_°C)$ relative to $C_6H_{14}O(l,\_\_\_°C)$.

  It is $\Delta\hat{H}$ for the process $C_6H_{14}O(l,\_\_\_°C) \to C_6H_{14}O(l,\_\_\_°C)$.

- $\hat{H}_3$ is the specific enthalpy of $C_3H_6O(l,\_\_\_\_\_)$ relative to $C_3H_6O(l,\_\_\_\_\_)$.

  It is $\Delta\hat{H}$ for the process $C_3H_6O(l,\_\_\_\_\_) \to C_3H_6O(l,\_\_\_\_\_)$.

Remember, $\hat{H}$ is always determined as $\Delta\hat{H}$ for the process in which the species goes from the reference state to the process state. Thus, even though in the process the methyl pentanol goes from 45°C to 20°C, when calculating $\hat{H}_2$ the opposite path must be used.

- $\hat{H}_2 = \int_{\_\_°C}^{\_\_°C} (C_p)_{C_6H_{14}O(l)}\, dT$ $\hat{H}_3 = \int_{\_\_°C}^{\_\_°C} (C_p)_{C_3H_6O(l)}\, dT$

The specific enthalpies $\hat{H}_1$ and $\hat{H}_4$ equal zero, because _____

We now need values or formulas for the heat capacities. The formula for C_p of liquid acetone is in Table B.2, but the one for 2-methyl-1-pentanol is not, and so we must either find it in another reference or estimate it. We'll do the latter, using *Kopp's rule* (the only heat capacity estimation formula given in the text—the formula is on p. 372 and the coefficients are given in Table B.10). First, the acetone.

(8.18-3)

$$(C_p)_{C_3H_6O(l)} = [0.1230 + _____ T]\ kJ/(mol \cdot °C)\ \ [\text{from Table B.2}]$$

$$\Rightarrow \hat{H}_3 = \int_{45}^{20}(0.1230 + _____ T\ )dT = 0.1230(\_\_ - \_\_) + \frac{\_\_\_\_ \times 10^{-5}}{2}(\_\_ - \_\_) = \_\_\_\_\_\ kJ/mol$$

$$= \frac{\_\_\_\_\_\ kJ}{mol}\ \left|\ \frac{1\ mol}{\_\_\_\_\ g}\ \right|\ \frac{\_\_\_\_\ g}{\_\_\_\_\ kg} = -\frac{\_\_\_\_\ kJ}{kg}$$

An alternative to integrating the polynomial heat capacity formula (which is not that lengthy a calculation in this case but can be tedious and a source of error when the polynomial contains four terms) is to go to *Interactive Chemical Process Principles,* open the Physical Property Database, click on the "Enthalpies" tab, select the species (liquid acetone) and enter the temperature limits, and determine the enthalpy change with a single click. Try it, and verify the calculated value of −3.226 kJ/mol.

Next, we estimate C_p for 2-methyl-1-pentanol using Kopp's rule (p. 372 in the text).

For 2-methyl-1-pentanol, we get:

(8.18-4)

$$\left(C_p\right)_{C_6H_{14}O(l)} \approx 6(C_{pa})_C + 14(C_{pa})_H + (C_{pa})_O$$

$$\xrightarrow{\text{Table B.10}} 6(\underline{\quad}) + 14(\underline{\quad}) + (\underline{\quad}) = \underline{\quad} \frac{kJ}{mol \cdot {}^\circ C}$$

$$\Rightarrow \hat{H}_2 = \int_{20}^{45} (\underline{\quad})dT = \underline{\quad}(45 - 20) = \frac{\underline{\quad} \, kJ}{mol} \left| \frac{1 \, mol}{\underline{\quad} \, g} \right| \frac{10^3 \, g}{1 \, kg} = \underline{\quad} \frac{kJ}{kg}$$

We can now fill in the enthalpy table.

(8.18-5)

References: $C_3H_6O(l,45^\circ C,1atm)$, $C_6H_{14}O(l,20^\circ C,1atm)$

	$\dot{n}_{in}$ (kg/min)	$\hat{H}_{in}$ (kJ/kg)	$\dot{n}_{out}$ (kg/min)	$\hat{H}_{out}$ (kJ/kg)
$C_3H_6O(l)$	____	____	____	____
$C_6H_{14}O(l)$	____	____	____	____

The final step is to solve the energy balance.

(8.18-6)

$$\dot{Q} = \Delta\dot{H} = \sum_{out}\dot{n}_i\hat{H}_i - \sum_{in}\dot{n}_i\hat{H}_i = (135)(-54.6) + (\underline{\quad})(0) - (\underline{\quad})(0) - (\underline{\quad})(\underline{\quad})$$

$$= \frac{\underline{\quad} \, kJ}{min} \left| \frac{1 \, min}{60 \, s} \right| \frac{1 \, kW}{1 \, kJ/s} = \underline{\quad} kW \, (\underline{\quad} kW \text{ of cooling})$$

One more thought before we leave this problem. For some reason that continues to baffle us, students often resist constructing enthalpy tables and try to keep track of all the problem quantities as a list instead. Invariably, this leads to errors. An enthalpy table is the best way to keep track of all the n's and H's and it makes applying the energy balance correctly MUCH easier. Do yourself a favor and take the extra few seconds it takes to construct an enthalpy table when one is needed.

Notes and Calculations

PROBLEM 8.22

A natural gas containing 95 mole% methane and the balance ethane is burned with 25% excess air. The stack gas leaves the furnace at 900°C and 1 atm and contains no unburned hydrocarbons or carbon monoxide. The gas is cooled to 450°C in a *waste heat boiler* (a heat exchanger in which heat lost by cooling gases is used to produce steam from liquid water for heating, power generation, or process applications) prior to being discharged into the atmosphere.

(a) Taking as a basis of calculation 100 mol of the fuel gas fed to the furnace, calculate the amount of heat (kJ) that must be transferred from the gas in the boiler to accomplish the indicated cooling.
(b) How much saturated steam at 50 bar can be produced from boiler feedwater at 40°C for the same basis of calculation? (Assume all the heat transferred from the gas goes into the steam production.)
(c) At what rate (kmol/s) must fuel gas be burned to produce 1250 kg steam per hour (an amount required elsewhere in the plant) in the waste heat boiler? What is the volumetric flow rate (m^3/h) of the gas leaving the boiler?
(d) Briefly explain how the waste heat boiler contributes to the plant profitability. (Think about what would be required in its absence.)

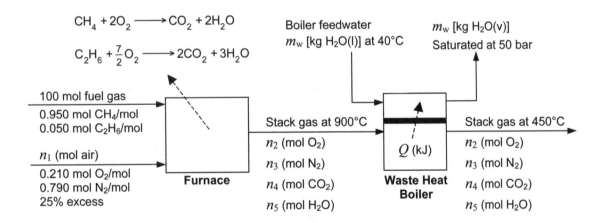

Strategy

Try to outline the approach you will use to solve Parts (a)–(c) of this problem. The first step and part of the second one are given for illustrative purposes.

(8.22-1)

- Do the degree-of-freedom analysis on the furnace to verify that the stack gas component amounts (n_2, n_3, n_4, n_5) can be determined. If they can, determine them.
- Write an energy balance on the gas side of the boiler to determine _____.
- _____

- _____

(a) Taking as a basis of calculation 100 mol of the fuel gas fed to the furnace, calculate the amount of heat (kJ) that must be transferred from the gas in the boiler to accomplish the indicated cooling.

Solution

(8.22-2)

DEGREE OF FREEDOM ANALYSIS — FURNACE		
UNKNOWNS AND INFORMATION		**JUSTIFICATION/CONCLUSION**
+ 5 unknowns	_____	
− __ balances	_____	Use atomic balances + N_2 balance
− 1 _____		
0 DOF		All unknowns can be determined

Exercise: Identify an order for the five system equations that would eliminate the need to solve any equations simultaneously.

(8.22-3)

(1) Write the excess air relation and solve it for n_1 (mol air fed)

(2) _____

(3) _____

(4) _____

(5) _____

25% excess air fed

theoretical oxygen

$$\left(n_1\right)(\text{mol air fed}) = \left[\frac{\_\_\ \text{mol CH}_4}{} \left|\frac{\_\_\ \text{mol O}_2}{1\ \text{mol CH}_4}\right. + \frac{\_\_\ \text{mol C}_2\text{H}_6}{} \left|\frac{\_\_\ \text{mol O}_2}{1\ \text{mol C}_2\text{H}_6}\right.\right]$$

$$\times \left[\frac{1\ \text{mol air}}{\_\_\ \text{mol O}_2}\right] \times \left[\frac{\text{mol air fed}}{\_\_\ \text{mol air theoretical}}\right]$$

(8.22-4)

N₂ balance _____

(8.22-5)

C balance

$$\frac{100\ \text{mol}}{}\left|\frac{\_\_\ \text{mol CH}_4}{\text{mol}}\right|\frac{\_\_\ \text{mol C}}{\_\_\ \text{mol CH}_4} + \frac{100\ \text{mol}}{}\left|\frac{\_\_\ \text{mol C}_2\text{H}_6}{\text{mol}}\right|\frac{\_\_\ \text{mol C}}{\_\_\ \text{mol C}_2\text{H}_6}$$

$$= \left(n_4\right)(\text{mol CO}_2)\left|\frac{\_\_\ \text{mol C}}{\_\_\ \text{mol CO}_2}\right.$$

(8.22-6)

___ balance _____

(8.22-7)

___ balance _____

(8.22-8)

These equations may be solved manually or with E-Z Solve to obtain

(8.22-9)

$n_1 = 1235$ mol air	$n_2 = $ _____ mol O_2 $n_3 = $ _____ mol N_2
$n_4 = $ _____ mol CO_2	$n_5 = 205$ mol H_2O

Next, we turn our attention to the energy balance for the gas side of the waste heat boiler. The flowchart of this part of the system appears as follows:

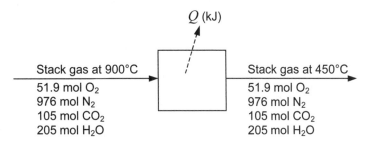

Even though we are labeling total amounts rather than flow rates, this is not a closed system; it is an open system for which we took a total amount of a stream as a basis. We could have chosen molar flow rates as well. We will therefore use the open-system energy balance equation. In preparation for calculating ΔH, we will set up an inlet-outlet enthalpy table.

We need four reference conditions to use for calculating the specific enthalpies of oxygen, nitrogen, carbon dioxide, and water vapor at 900°C and 450°C. If we planned to integrate C_p to do the calculations, we would choose either 900°C or 450°C as the reference so that half of the specific enthalpies to be calculated could be set equal to zero. Fortunately, we don't have to go to that much trouble, since in Table B.8 on p. 652 of the text there is an enthalpy table that includes all four substances. For convenience, then, we will choose the reference conditions used for Table B.8—the gases at 25°C and 1 atm. (As usual, we will neglect the effect of pressure on enthalpy and not worry about the pressure of the stack gas.)

(8.22-9)

References: O_2, N_2, CO_2, $H_2O(v)$ at 25°C and 1 atm

	n_{in} (mol)	$\hat{H}_{in}$ (kJ/mol)	n_{out} (mol)	$\hat{H}_{out}$ (kJ/mol)
O_2	51.9	28.89	51.9	13.38
N_2	____	____	____	____
CO_2	____	____	____	____
$H_2O(v)$	____	____	____	____

After neglecting (as usual) shaft work and kinetic and potential energy changes, the open system balance equations (without dots over the variables, since we are working with amounts and not flow rates) becomes

$$Q = \Delta H = \sum_{out} n_i \hat{H}_i - \sum_{in} n_i \hat{H}_i = (51.9 \text{ mol } O_2)\left[(13.38 - 28.89)\frac{kJ}{\text{mol } O_2}\right]$$

(8.22-10)

$$+976(12.70 - 27.19) + \text{_____} = \text{_____} \text{ kJ}$$

(b) How much saturated steam at 50 bar can be produced from boiler feedwater at 40°C for the same basis of calculation? (Assume all the heat transferred from the gas is used to produce steam.)

Solution

An energy balance on the steam side of the boiler (for which we now know Q) will have only the amount of the steam as an unknown variable. Here is the flowchart.

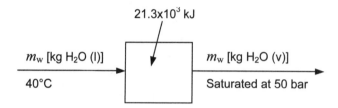

21.3x10³ kJ

m_w [kg H₂O (l)]

40°C

m_w [kg H₂O (v)]

Saturated at 50 bar

Energy balance

$$Q = \Delta H \Rightarrow 21.3 \times 10^3 \text{ kJ} = m_w[\hat{H}_{H_2O(v,\text{sat'd},50\text{ bar})} - \hat{H}_{H_2O(l,40°C)}]$$

$$\xrightarrow{\text{Tables B.5 and B.6}} 21.3 \times 10^3 \text{ kJ} = m_w(\text{kg})\left[(\underline{\hspace{1cm}} - \underline{\hspace{1cm}})\frac{\text{kJ}}{\text{kg}}\right] \qquad \text{(8.22-11)}$$

$$\Rightarrow m_w = \underline{\hspace{1cm}} \text{ kg steam produced}$$

(c) At what rate (kmol/s) must fuel gas be burned to produce 1250 kg steam per hour (an amount required elsewhere in the plant) in the waste heat boiler?

Solution

To scale the process up to the desired steam production rate, we must multiply all stream amounts corresponding to the original basis by

$$S = \frac{\underline{\hspace{1cm}} \text{ kg steam/h}}{\underline{\hspace{1cm}} \text{ kg steam}} = 154.1 \text{ h}^{-1}$$

If we multiply this scale factor times the fuel feed quantity for the original basis we obtain

$$(\underline{\hspace{1cm}} \text{ mol fuel}) \times (154 \text{ h}^{-1}) = \frac{\underline{\hspace{1cm}} \text{ mol}}{\text{h}} \left| \frac{1 \text{ kmol}}{10^3 \text{ mol}} \right| \frac{1 \text{ h}}{3600 \text{ s}} \qquad \text{(8.22-12)}$$

$$= \underline{\hspace{1cm}} \text{ kmol fuel/s}$$

We'll use the ideal gas equation of state to determine the volumetric flow rate of the exiting stack gas, first scaling up the total molar flow rate under the original basis and assuming that the stack gas is at 1 atm. The scaled-up total molar flow rate of the stack gas is

$$\dot{n}_{\text{stack}} = (\underline{\hspace{0.5cm}} + \underline{\hspace{0.5cm}} + \underline{\hspace{0.5cm}} + \underline{\hspace{0.5cm}}) \text{ mol} \times 154.1 \text{ h}^{-1} \times \left(\frac{1 \text{ kmol}}{10^3 \text{ mol}}\right) = 206.2 \frac{\text{kmol}}{\text{h}} \qquad \text{(8.22-13)}$$

$$\dot{V}_{\text{stack}} = \frac{206.2 \text{ kmol}}{\text{h}} \left| \frac{\underline{\hspace{1cm}} \text{m}^3(\text{STP})}{\text{kmol}} \right| \frac{(\underline{\hspace{1cm}})\text{K}}{273 \text{ K}} \left| \frac{1 \text{ h}}{3600 \text{ s}} \right| = \underline{\hspace{1cm}} \text{ m}^3/\text{s} \qquad \text{(8.22-14)}$$

(d) Briefly explain how the waste heat boiler contributes to the plant profitability. (Think about what would be required if there were no boiler.)

Solution   (8.22-15)

Notes and Calculations

PROBLEM 8.30

A stream of humid air at 500°C and 835 torr with a dew point of 30°C flowing at a rate of 1515 liters/s is to be cooled in a spray cooler. A fine mist of liquid water at 25°C is sprayed into the hot air at a rate of 110.0 g/s and evaporates completely. The cooled air emerges at 1 atm.

(a) Calculate the final temperature of the emerging air stream, assuming that the process is adiabatic. (*Suggestion:* Derive expressions for the enthalpies of dry air and water at the outlet air temperature, substitute them into the energy balance, and use E-Z Solve to solve the resulting fourth-order polynomial equation.)
(b) At what rate (kW) is heat transferred from the hot air feed stream in the spray cooler? What becomes of this heat?
(c) In a few sentences, explain how this process works in terms that a high school senior could understand. Incorporate the results of Parts (a) and (b) in your explanation.

How does this process work?

In this *spray cooler*, liquid water evaporates into hot air at a temperature well above the moistened air's dew point. The energy required for the vaporization has to come from someplace. Since no heat is being supplied to the system externally, the only place it can come from is the total internal energy of the gas, which drops accordingly. As the internal energy decreases, the temperature (which is directly related to the internal energy) also drops. Besides being the basis of spray cooling, the temperature drop that accompanies evaporation with no external heating explains why you feel cold when you step out of a shower and why on a hot day you feel cooler if you stand near a spraying fountain.

(a) Here is the flowchart of the process.

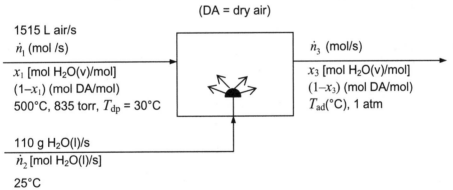

(DA = dry air)

1515 L air/s
$\dot{n}_1$ (mol /s)
x_1 [mol H_2O(v)/mol]
$(1-x_1)$ (mol DA/mol)
500°C, 835 torr, T_{dp} = 30°C

110 g H_2O(l)/s
$\dot{n}_2$ [mol H_2O(l)/s]
25°C

$\dot{n}_3$ (mol/s)
x_3 [mol H_2O(v)/mol]
$(1-x_3)$ (mol DA/mol)
T_{ad}(°C), 1 atm

Strategy

- Do the degree-of-freedom analysis and verify that there is enough information to determine the final temperature.
- Write the relevant density, gas law, and phase equilibrium equations and the system material balance equations.
- Calculate the specific enthalpies of the inlet stream components, and express the enthalpies of the outlet water vapor and dry air in terms of T_{ad} (by integrating their heat capacities from a reference temperature to T_{ad}).
- Write and solve the energy balance equation for T_{ad}. [Solves Part (a).]
- Calculate the cooling required to bring the inlet air from its initial condition to the (now known) T_{ad}. [Solves the quantitative part of Part (b).]
- Calculate the final temperature of the emerging air stream, assuming that the process is adiabatic.

Name: _____

Date: _____

Solution

(8.30-1)

DEGREE OF FREEDOM ANALYSIS		
UNKNOWNS AND INFORMATION		JUSTIFICATION/CONCLUSION
+ 6 unknowns	$\dot{n}_1\text{-}\dot{n}_3, x_1, x_2, T_{ad}$	
− 2 balances		
− 1 gas law at inlet	Calculate	
− 1 inlet air dew point	Calculate	
− 1 liquid density	Calculate	
− 1 energy balance		
0 DOF		All unknowns can be determined

Rather than doing the algebra and arithmetic manually, we will simply write the equations and use E-Z Solve to solve them all simultaneously. (Manual solution of the energy balance would require a trial-and-error solution of a fourth-order polynomial equation.)

Ideal gas equation of state for inlet gas

$$\dot{n}_1 = \frac{P_1 \dot{V}_1}{RT} = \frac{\underline{\quad} \text{ torr}}{\underline{\quad} \text{ K}} \bigg| \frac{\underline{\quad} \text{ L}}{\text{s}} \bigg| \frac{\text{mol} \cdot \text{K}}{\underline{\quad} \text{ L} \cdot \text{torr}}$$ (8.30-2)

Raoult's law for inlet gas (Eq. 6.3-3, p. 250 of the text)

$$x_1 P = p_{H_2O}^* (\underline{\quad} °C) \Rightarrow x_1 (\underline{\quad} \text{ torr}) = \underline{\quad} \text{ torr}$$ (8.30–3)

Table B.3

Mass flow rate of liquid water feed

$$\dot{n}_2 = \left(\underline{\quad} \frac{\text{g}}{\text{s}} \right) \left(\frac{1 \text{ mol}}{18.02 \text{ g}} \right)$$ (8.30-4)

Total mole balance

$$\dot{n}_1 + \dot{n}_2 = \underline{\quad}$$ (8.30-5)

Water balance

_____ (8.30-6)

We could stop and solve **(8.30-2)**–**(8.30-5)** for the three flow rates and two mole fractions, but instead we'll continue and solve the whole set of system equations at the same time. Let us proceed to the specific enthalpies, which we will need to substitute into the energy balance (which will reduce to $\Delta \dot{H} = 0$ for this adiabatic system). As usual, we will use an inlet-outlet enthalpy table to help keep track of the calculations.

First, we need to decide what to use as reference conditions for our two species (dry air and water). The choice depends on how the enthalpies will be calculated. In the problems we've worked so far, we knew the temperatures of the process inlet and outlet streams and could use the steam tables for water (we could also use Table B.8 for water vapor and dry air). In this problem, however, we don't know the outlet temperature and so cannot look up the outlet enthalpies in tables.

Instead, we will have to integrate the heat capacity formula up to the unknown T_{ad} for each enthalpy needed, substitute the enthalpy expressions into the energy balance, and solve the resulting fourth order polynomial for T_{ad} (thank goodness for E-Z Solve!). That being the case, we may as well use the inlet air condition (500°C and 1 atm) as the references for both species so we can use $\hat{H} = 0$ for the dry air and water vapor at the inlet.

(8.30-7)

References: Dry air, $H_2O(v)$, 500°C and 1 atm				
	$\dot{n}_{in}$ (mol/s)	$\hat{H}_{in}$ (kJ/mol)	$\dot{n}_{out}$ (mol/s)	$\hat{H}_{out}$ (kJ/mol)
Dry air	$\dot{n}_1(1-x_1)$	_____	_____	$\hat{H}_b$
H_2O (v)	_____	_____	_____	$\hat{H}_c$
H_2O (l)	_____	$\hat{H}_a$		

Specific enthalpy of liquid feed

We need to calculate $\hat{H}_a$ as $\Delta\hat{H}$ for a process in which water goes from vapor at 500°C and 1 atm (the reference state) to liquid at 25°C (the process state). We can use the steam tables for this calculation since both the initial and final states are known.

$H_2O(v, 500°C, 1\ atm) \rightarrow H_2O(l, 25°C)$:

$$\hat{H}_a = \Delta\hat{H} = \hat{H}_{H_2O(l,\,25°C)} - \hat{H}_{H_2O(v,\,500°C,\,1\ atm)}$$

(8.30-8)

$$= (\_\_\_\_ - \_\_\_\_)\frac{kJ}{kg} \times \left(\frac{18.02\ kg}{kmol}\right) \times \left(\frac{1\ kmol}{10^3\ mol}\right)$$

Specific enthalpy of outlet air (Use Table B.2 for the heat capacity.)

$Air(g, 500°C, 1\ atm) \rightarrow Air\ (g, T_{ad}, 1\ atm)$:

$$\hat{H}_b = \Delta\hat{H} = \int_{500}^{T_{ad}} (C_p)_{air}\,dT$$

$$= \int_{500}^{T_{ad}} \left(.02894 + 0.4147 \times 10^{-5}T + _____T^2 - _____T^3\right)dT$$

(8.30-9)

$$= 0.02894(T_{ad} - 500) + \frac{0.4147 \times 10^{-5}}{2}\left(T_{ad}^2 - 500^2\right)$$

$$+ \frac{\_\_\_\_ \times 10^{-8}}{\_\_\_\_}\left(T_{ad}^3 - 500^3\right) - \frac{\_\_\_\_ \times 10^{-12}}{\_\_\_\_}\left(T_{ad}^4 - 500^4\right)$$

Specific enthalpy of outlet water vapor. (Use Table B.2 for the heat capacity.)

$H_2O(g, 500°C, 1\ atm) \rightarrow H_2O\ (g, T_{ad}, 1\ atm)$:

$$\hat{H}_c = \Delta\hat{H} = \int_{500}^{T_{ad}} (C_p)_{H_2O(v)}\,dT$$

$$= \int_{500}^{T_{ad}} \left(.03346 + 0.6880 \times 10^{-5}T + 0.7604 \times 10^{-8}T^2 - 3.593 \times 10^{-12}T^3\right)dT$$

(8.30-10)

$$= 0.03346\ _____ + \frac{0.6880 \times 10^{-5}}{2}\ _____$$

$$+ \frac{0.7604 \times 10^{-8}}{3}\ _____ - \frac{3.593 \times 10^{-12}}{4}\ _____$$

Energy balance

$$\dot{Q} - \dot{W}_s = \Delta\dot{H} + \Delta\dot{E}_k + \Delta\dot{E}_p$$

$$\begin{cases} \dot{Q} = 0 & \text{(system is adiabatic)} \\ \dot{W}_s = 0 & \text{(no moving parts)} \\ \Delta\dot{E}_k = 0 & \left(\text{neglect kinetic energy change from inlet to outlet}\right) \\ \Delta\dot{E}_p = 0 & \left(\text{neglect vertical displacement between inlet and outlet}\right) \end{cases}$$

$$0 = \Delta\dot{H} = \sum_{\text{out}} \dot{n}_i \hat{H}_i - \sum_{\text{in}} \dot{n}_i \hat{H}_i$$

From the enthalpy table, this equation becomes

$$0 = \dot{n}_3(1-x_3)\hat{H}_b + \underline{\hspace{3cm}}$$ (8.30-11)

E-Z Solve program   (8.30-12)

```
//Problem 8.30

//Flow rates, Raoult's law, and material balances

n1 = 835*1515/(773*62.36)        // Ideal gas EOS for inlet air

x1*835 = 31.82                    // Raoult's law for inlet air

n2 = _____            // Conversion of mass to molar flow rate for water feed

_____                 // Mole balance

_____                 // Water balance

//Enthalpies of air and water at process conditions

Ha = (104.8-3488)*18.02/1000

Hb = 0.02894*(T–500+0.4147e–5*(T^2–500^2)/2+0.3191e–8*(T^3–500^3)/3–1.965e–12*(T^4–500^4)/4

Hc = 0.03346*(T–500)+0.6880e–5*(T^2–500^2)/2+0.7604e–8*(T^3–500^3)/3–3.593e–12*(T^4–500^4)/4

//Energy balance

n3*(1-x3)*Hb + _____ = 0
```

E-Z Solve solution   (8.30-13)

$$n_1 = 26.2 \text{ mol/s} \qquad x_1 = 0.0381 \text{ mol } H_2O(v)/\text{mol} \qquad n_2 = \underline{\hspace{2cm}}$$

$$n_3 = \underline{\hspace{2cm}} \qquad x_3 = \underline{\hspace{3cm}} \qquad T_{\text{ad}} = \underline{\hspace{2cm}}$$

(b) At what rate (kW) is heat transferred from the hot air feed stream in the spray cooler? What becomes of this heat?

Solution

Consider a process in which the entering air is cooled from its initial state to the state of the outlet air in Part (a)—138°C and 1 atm.

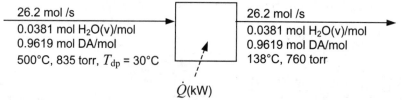

26.2 mol /s
0.0381 mol $H_2O(v)$/mol
0.9619 mol DA/mol
500°C, 835 torr, $T_{\text{dp}} = 30$°C

26.2 mol /s
0.0381 mol $H_2O(v)$/mol
0.9619 mol DA/mol
138°C, 760 torr

$\dot{Q}$(kW)

Name: ____________________

Date: ____________

The energy balance for this process is

Table B.8

$$\dot{Q} = \Delta\dot{H} = \sum_{out}\dot{n}_i\hat{H}_i - \sum_{in}\dot{n}_i\hat{H}_i = \dot{n}_{DA}\left[\hat{H}_{DA(138°C)} - \hat{H}_{DA(500°C)}\right] + \dot{n}_{H_2O}\left[\hat{H}_{H_2O(v,138°C)} - \hat{H}_{H_2O(v,500°C)}\right]$$

$$= \left(26.2 \times 0.9619\ \frac{\text{mol DA}}{\text{s}}\right)\left[\left(______ - ______\right)\frac{\text{kJ}}{\text{mol DA}}\right]$$

$$+\left(26.2 \times 0.0381\ \frac{\text{mol H}_2\text{O(v)}}{\text{s}}\right)\left[\left(______ - ______\right)\frac{\text{kJ}}{\text{mol H}_2\text{O(v)}}\right] \qquad \textbf{(8.30-14)}$$

$$= \frac{____\ \text{kJ}}{\text{s}}\left|\frac{1\ \text{kW}}{1\ \text{kJ/s}}\right. = \underline{\underline{____\ \text{kW}}}$$

This energy goes to ____. **(8.30-15)**

(c) In a few sentences, explain how this process works in terms that a high school senior could understand. Incorporate the results of Parts (a) and (b) in your explanation.

(8.30-16)

Notes and Calculations

PROBLEM 8.46

Humid air at 50°C and 1.0 atm with 2°C of superheat is fed to a condenser. Gas and liquid streams leave the condenser in equilibrium at 20°C and 1 atm.

(a) Assume a basis of calculation of 100 mol inlet air, draw and label a flowchart [including $Q(kJ)$ in the labeling], and carry out a degree-of-freedom analysis to verify that all labeled variables can be determined.

(b) Write the equations you would solve to calculate the mass of water condensed (kg) per cubic meter of air fed to the condenser. Do not do any of the calculations.

(c) Prepare an inlet-outlet enthalpy table, inserting labels for unknown specific enthalpies. Write expressions for the labeled specific enthalpies and then write the other equations you would use to calculate the rate at which heat must be transferred from the unit (kJ) per cubic meter of air fed to the condenser.

(d) Solve your equations to calculate kg H_2O condensed/m^3 air fed and kJ transferred/m^3 air fed. What cooling rate (kW) would be required to process 250 m^3 air fed/h?

The elements of the problem solution should all be familiar to you by now, so we will go through them without much additional explanation.

(a) Assume a basis of calculation of 100 mol humid inlet air, draw and label a flowchart [including $Q(kJ)$ in the labeling], and carry out a degree-of-freedom analysis to verify that all labeled variables can be determined.

Solution

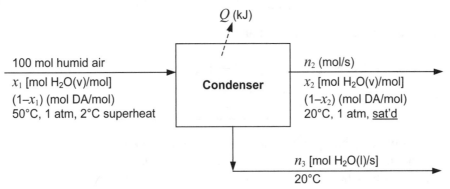

(8.46-1)

Q: We noted on the chart that the water vapor in the exit air is saturated. How do we know?

A: _____

(8.46-2)

DEGREE OF FREEDOM ANALYSIS		
UNKNOWNS AND INFORMATION		**JUSTIFICATION/CONCLUSION**
+__ unknowns	_____	
−__ material balances	_____	
− 1 _____		Calculate ____
− 1 _____		Calculate ____
− 1 energy balance		
___ DOF		_____

Name: _____

Date: _____

(b) Write the equations you would solve to calculate the mass of water condensed (kg) per cubic meter of air fed to the condenser. Do not do any of the calculations.

Solution

(8.46-3)

Raoult's law for inlet air (Eq. 6.3-3, p. 250 in the text)

$$T_{dp} = \underline{\qquad} °C \Rightarrow x_1 P = p_{H_2O}^* (\underline{\qquad}°C) \Rightarrow \boxed{x_1}(760 \text{ torr}) = \underline{\qquad} \text{ torr}$$

Table B.3

Raoult's law for outlet air

$$x_2 P = \underline{\qquad} \Rightarrow \boxed{x_2}(760 \text{ torr}) = \underline{\qquad} \text{ torr}$$

Dry air balance

$$\underline{\qquad} = \boxed{n_2} \times \underline{\qquad}$$

Water balance

$$\underline{\qquad} = \underline{\qquad} + \boxed{n_3}$$

Volume of air fed

$$\boxed{V_f}(\text{m}^3) = \frac{100 \text{ mol air fed}}{} \left| \frac{\underline{\qquad} \text{ L(STP)}}{\text{mol}} \right| \frac{\underline{\qquad} \text{ K}}{\underline{\qquad}\text{ K}} \left| \frac{1 \text{ m}^3}{10^3 \text{ L}} \right.$$

Mass of water condensed per cubic meter of air fed

$$\boxed{m_c / V_f} = \frac{n_3(\text{mol H}_2\text{O condensed})] \times \left[0.018 \dfrac{\text{kg}}{\text{mol}} \right]}{V_f(\text{m}^3)}$$

(c) Prepare an inlet-outlet enthalpy table, inserting labels for unknown specific enthalpies. Write expressions for the labeled specific enthalpies and then write the other equations you would use to calculate the rate at which heat must be transferred from the unit (kJ) per cubic meter of air fed to the condenser.

Solution

We choose reference states for water and dry air based on how we plan to evaluate their specific enthalpies at the inlet and outlet.

(8.46-4)

Q: What reference states are most convenient, and why?

A: Since both inlet and outlet temperatures are known, we can use Table _____ for dry air, and so would choose dry air (_____°C, 1 atm) as a reference state (the state used for that table). That table cannot be used for liquid water, and so instead we will use the steam tables, for which the reference is water (l, _____°C, _____ bar). (*Hint:* $\hat{H} = 0$ at the reference point.)

With the reference states chosen, we can construct the inlet-outlet enthalpy table.

(8.46-5)

References: Dry air (___ °C, 1 atm), H_2O (l, ___ °C, _____ bar)

	n_{in} (mol)	$\hat{H}_{in}$ (kJ/mol)	n_{out} (mol)	$\hat{H}_{out}$ (kJ/mol)
Dry air	$100(1-x_1)$	$\hat{H}_{1a}$	_____	$\hat{H}_{2a}$
H_2O (v)	_____	$\hat{H}_{1w}$	_____	$\hat{H}_{2w}$
H_2O (l)			_____	$\hat{H}_{3w}$

As usual, we will neglect the effect of pressure on specific enthalpy and use the tabulated enthalpies for air at 1 atm in Table B.8 and for saturated liquid and water vapor in Table B.5.

(8.46-6)

Dry air (50°C) $\hat{H}_{1a}$ = _____ kJ/mol   **Dry air (20°C)** $\hat{H}_{2a}$ = _____ kJ/mol

H_2O (v, 50°C) $\hat{H}_{1w}$ = $\dfrac{\text{——} \ \text{kJ}}{\text{kg}} \left| \dfrac{0.01802 \ \text{kg}}{1 \ \text{mol}} \right.$ **H_2O (v, 20°C)** $\hat{H}_{2w}$ = $\dfrac{\text{——} \ \text{kJ}}{\text{kg}} \left| \dfrac{0.01802 \ \text{kg}}{1 \ \text{mol}} \right.$

H_2O (l, 20°C) $\hat{H}_{3w}$ = $\dfrac{\text{——} \ \text{kJ}}{\text{kg}} \left| \dfrac{0.01802 \ \text{kg}}{1 \ \text{mol}} \right.$

Energy balance (We've neglected shaft work and kinetic and potential energy changes.)

$$\boxed{Q} = \Delta H = \sum_{out} n_i \hat{H}_i - \sum_{in} n_i \hat{H}_i \quad \text{(substitute from enthalpy table)}$$

(8.46-7)

$$= n_2(1-x_2)H_{2a} + \underline{\hspace{8cm}}$$

Energy transferred per cubic meter of air fed = $\boxed{Q/V_f}$

(d) Solve your equations to calculate kg H_2O condensed/m³ air fed and kJ transferred/m³ air fed.

E-Z Solve program

```
//E-Z Solve program for Problem 8.46

x1*760 = _____          // Raoult's law for inlet air

x2*760 = _____          // Raoult's law for outlet air

_____ = n2*_____          // Dry air balance

_____ = _____ + n3          // Water balance
Vf = 100*22.4*(323/273)/1000          // Volume of air fed
M_V = n3*0.018/Vf          // kg water condensed per cubic meter of air fed
H1a = _____          // specific enthalpy of dry air at inlet

H2a = _____          // specific enthalpy of dry air at outlet

H1w = _____*0.01802          // specific enthalpy of water vapor at inlet

H2w = _____*0.01802          // specific enthalpy of water vapor at outlet

H3w = _____*0.01802          // specific enthalpy of liquid water at outlet

Q = n2*(1–x2)*H2a + n2*x2*H2w + n3*H3w – 100*(1–x1)*H1a – 100*x1*H1w          // energy balance
Q_V = Q/Vf          // kJ transferred per cubic meter of air fed
```

Name: _____

Date: _____

E-Z Solve solution (8.46-8)

Unknown variables

$x_1 =$ _____ mol $H_2O(v)$/mol $n_2 =$ _____ mol

$x_2 =$ _____ mol $H_2O(v)$/mol $n_3 =$ _____ mol $H_2O(l)$

$V_f =$ _____ m^3 air fed $Q =$ _____ kJ

Ratios requested

_____ kg H_2O condensed/m^3 air fed   & -182 kJ/m^3 air fed

(e) What cooling rate (kW) would be required to process 250 m^3 air fed/h?

Solution

The ratios just calculated are valid regardless of the basis of calculation. The cooling rate is therefore easily calculated by using the energy ratio as a scale factor.

$$\dot{Q} = \frac{\underline{\quad\quad}\ m^3\ \text{air fed}}{h} \left| \frac{\underline{\quad\quad}\ kJ}{m^3} \right| \frac{1\ h}{3600\ s} \left| \frac{1\ kW}{1\ kJ/s} \right. = -\underline{\quad\quad}\ kW \Rightarrow \underline{\quad\quad}\ kW\ \text{of cooling} \quad (8.46\text{-}9)$$

PROBLEM 8.52

A liquid stream containing 50.0 mole% benzene and the balance toluene at 25.0°C is fed to a continuous single-stage evaporator at a rate of 1320 mol/s. The liquid and vapor streams leaving the evaporator are both at 95.0°C. Both streams are analyzed and are found to contain 42.5 mole% benzene and 73.5 mole% benzene, respectively.

(a) Calculate the heating requirement for this process in kW.

Solution

We approach this problem like all the others we have done—flowchart, degree-of-freedom analysis, material balances for molar flow rates and compositions, and then the energy balance to determine $\dot{Q}$.

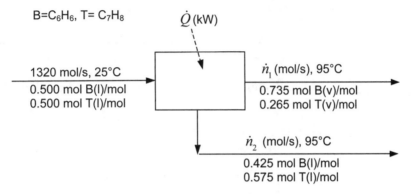

B=C_6H_6, T= C_7H_8 $\dot{Q}$ (kW)

1320 mol/s, 25°C
0.500 mol B(l)/mol
0.500 mol T(l)/mol

$\dot{n}_1$ (mol/s), 95°C
0.735 mol B(v)/mol
0.265 mol T(v)/mol

$\dot{n}_2$ (mol/s), 95°C
0.425 mol B(l)/mol
0.575 mol T(l)/mol

(8.52-1)

DEGREE OF FREEDOM ANALYSIS		
UNKNOWNS AND INFORMATION		**JUSTIFICATION/CONCLUSION**
+ 3 unknowns	$\dot{n}_1, \dot{n}_2, \dot{Q}$	
− __ material balances	_____	
− __ energy balance		
0 DOF		All unknowns can be determined

(8.52-2)

Mole balance _____ $\dot{n}_1 = 319.4$ mol/s

$\Rightarrow$

Benzene balance _____ $\dot{n}_2 = $ _____ mol/s

Next, prepare the inlet-outlet enthalpy table.

(8.52-3)

Q: Not counting the reference states and column heading row, how many rows will the table need, and why?

A: _____

(8.52-4)

Q: Since we are not going to use tabulated enthalpy data, the choice of reference states is arbitrary. What are convenient choices for benzene? What makes them convenient?
A: _____
Q: From Table B.1, the normal boiling point of benzene is 80.1°C. How can benzene exist as a liquid at 95°C and 1 atm as it does in this process?
A: _____

Here is the enthalpy table. We'll choose the liquids of each species at 95°C and 1 atm as references.

(8.52-5)

References: B(l, 95°C, 1 atm), T(l, 95°C, 1 atm)

	$\dot{n}_{in}$ (mol)	$\hat{H}_{in}$ (kJ/mol)	$\dot{n}_{out}$ (mol)	$\hat{H}_{out}$ (kJ/mol)
B(l)	660	$\hat{H}_1$	_____	_____
T(l)	_____	$\hat{H}_2$	_____	_____
B(v)			(319.4)(0.735)	$\hat{H}_3$
T(v)			_____	$\hat{H}_4$

To determine each of the four specific enthalpies listed in the table, we must calculate $\Delta\hat{H}$ for the process in which the species in question goes from its reference state (liquid at 95°C) to the process condition (liquid at 25°C for $\hat{H}_1$ and $\hat{H}_2$, vapor at 95°C for $\hat{H}_3$ and $\hat{H}_4$), constructing a path that allows us to use the physical property data in the text. Let's do that next.

Benzene in feed

$$B(l, \underline{\quad}°C) \rightarrow B(l, \underline{\quad}°C) \qquad \Delta\hat{H} = \hat{H}_1 = \int_{\underline{\quad}°C}^{\underline{\quad}°C} (C_p)_{B(l)} dT = -9.838 \text{ kJ/mol}$$ (8.52-6)

We could have calculated that value by looking up the heat capacity of liquid benzene in Table B.2 and integrating it between the two limits (not worrying about the fact that 95°C is slightly out of the listed range of applicability of the heat capacity formula—it will barely affect the result). What we actually did was go to the Physical Property Database in *Interactive Chemical Process Principles*, clicked on the "Enthalpies" tab, selected liquid benzene and the two temperature limits, and let the computer do the rest.

Toluene in feed

$$T(l, \underline{\quad}°C) \rightarrow T(l, \underline{\quad}°C) \qquad \Delta\hat{H} = \hat{H}_2 = \int_{\underline{\quad}°C}^{\underline{\quad}°C} (C_p)_{\underline{\quad}} dT = \underline{\qquad} \text{ kJ/mol}$$ (8.52-7)

Benzene vapor at outlet   (8.52-8)

> **Q:** $\hat{H}_3$ is the specific enthalpy change for the process in which liquid benzene at 95°C goes to benzene vapor at 95°C. By definition, this is what physical property of benzene?
>
> **A:** _____ .
>
> **Q:** Why can't we just look up this value in the text and enter it in the inlet-outlet enthalpy table? Since we can't, what do we need to do to calculate $\hat{H}_3$?
>
> **A:** _____
> _____

B(l, 95°C) → B(l, 80.1°C) → B(v, 80.1°C) → B(v, 95°C)

$$\Delta\hat{H} = \hat{H}_3 = \int_{95°C}^{\underline{\quad}°C} \left(C_p\right)_{B(l)} dT + \left(\Delta\hat{H}_v\right)_{B(\underline{\quad}°C)} + \int_{\underline{\quad}°C}^{95°C} \left(C_p\right)_{B(v)}$$

                                                                                                            (8.52-9)

$$= -\underline{\quad}\frac{kJ}{mol} + \underline{\quad}\frac{kJ}{mol} + \underline{\quad}\frac{kJ}{mol} = \underline{\quad}\frac{kJ}{mol}$$

Incidentally, we now know that the heat of vaporization of benzene at 95°C is 30.08 kJ/mol.

Toluene vapor at outlet

T(l, 95°C) → T (l, _____°C) → T (v, _____°C) → T (v, 95°C)

$$\Delta\hat{H} = \hat{H}_4 = \underline{\qquad\quad} + \underline{\qquad\quad} + \underline{\qquad\quad}$$

                                                                                                            (8.52-10)

$$= \underline{\quad}\frac{kJ}{mol} + \underline{\quad}\frac{kJ}{mol} - \underline{\quad}\frac{kJ}{mol} = \underline{\quad}\frac{kJ}{mol}$$

The enthalpy table may now be filled in completely.

References: B(l, 95°C, 1 atm), T(l, 95°C, 1 atm)

	$\dot{n}_{in}$ (mol)	$\hat{H}_{in}$ (kJ/mol)	$\dot{n}_{out}$ (mol)	$\hat{H}_{out}$ (kJ/mol)
B(l)	660	−9.838	(1000.6)(0.425)	0
T(l)	660	-11.78	(1000.6)(0.575)	0
B(v)			(319.4)(0.735)	30.08
T(v)			(319.4)(0.265)	34.28

Energy balance (We've neglected shaft work and kinetic and potential energy changes.)

$$\dot{Q} = \Delta\dot{H} = \sum_{out} \dot{n}_i \hat{H}_i - \sum_{in} \dot{n}_i \hat{H}_i \quad \text{(substitute from the enthalpy table)}$$

                                                                                                            (8.52-11)

$$= (\underline{\qquad\quad} kJ/s)\left(\frac{\underline{\quad} kW}{\underline{\quad} kJ/s}\right) = \underline{\qquad} kW$$

(b) Using Raoult's law (Section 6.4b in the text) to describe the equilibrium between the vapor and liquid outlet streams, determine whether or not the given benzene analyses are consistent with each other. If they are, calculate the evaporator's operating pressure (torr); if they are not, give several possible explanations for the inconsistency.

Solution

Let us write Raoult's law (Eq. 6.4-1 on p. 257 of the text) for benzene and toluene, using the Antoine equation (Table B.4) to determine the pure-species vapor pressures at 95°C. For ease of reference, here again are the given phase compositions.

Vapor	73.5% B	26.5% T
Liquid	42.5% B	57.5% T

Vapor pressures From Table B.4, $p_B^*(95°C) = 1177$ torr , $p_T^*(95°C) = 476.9$ torr

Raoult's law

> **Benzene** $p_B = x_B p_B^*(95°C) =$ (_____)(_____) = 500 torr (8.52-12)

> **Toluene** $p_T =$ _____ = (_____)(_____) = _____ torr (8.52-13)

and the mole fraction of benzene in the vapor should be

$$y_B = \frac{p_B}{P} = \frac{p_B}{p_B + p_T} = \frac{500 \text{ torr}}{\rule{1cm}{0.4pt} \text{ torr}} = \frac{\text{mol B(v)}}{\text{mol}}$$ (8.52-14)

This value is significantly different from the reported value of 0.735, leading to the possible conclusion that the reported molar percentages of benzene in the liquid and vapor are <u>inconsistent</u> with each other. List some possible reasons for the inconsistency below.

(8.52-15)

PROBLEM 8.72

Humid air is enclosed in a 2.00-liter flask at 40°C. The flask is slowly cooled. When the temperature reaches 20°C, drops of moisture become visible on the flask wall. Although the pressure in the flask changes when the temperature drops, it remains close enough to 1 atm for the psychrometric chart to provide a close representation of the behavior of the system throughout the process. Use the chart to solve the following problems.

(a) What were the relative humidity, absolute humidity, wet-bulb temperature, and humid volume of the air at 40°C?

(b) Calculate the mass of the water in the flask.

(c) Calculate the enthalpy change (J) of the air as it cools from 40°C to 20°C.

(d) Write an energy balance for this closed-system process, taking the wet air in the flask as the system, and use it to calculate the heat in joules that must be transferred from the air to accomplish the cooling. (Assume ideal gas behavior, so that $\hat{H} = \hat{U} + P\hat{V} = \hat{U} + RT$.)

Strategy

There are a number of properties of humid air that are either directly plotted or can easily be determined in the psychrometric chart, including dry-bulb and wet-bulb temperatures, relative and absolute humidity, dew point, humid volume, and enthalpy. If we have values of two of them for the air at its initial condition, we can locate the position of the air on the chart and then determine the values of the other properties. We can also trace the path of the air as it cools to its final condition, determine all the properties of the air at that condition, and then carry out the requested process calculations. Note that what you gain in convenience when using the psychrometric chart you lose in accuracy, and so the answers you get may deviate slightly from ours.

(a) What were the relative humidity, absolute humidity, and wet-bulb temperature of the air at 40°C?

Solution

(8.72-1)

Q: The temperature of the gas is given as 40°C. Is that the wet-bulb or dry-bulb temperature?
A: _____.
Q: What other property of the air that can be located on the psychrometric chart is given in the process statement?
A: We are told that if the gas is cooled at (essentially) constant pressure, condensation begins at 20°C. By definition, the _____ = 20°C.
Q: How do we use that information to locate the point corresponding to the initial air? (Refer to Fig. 8.4-1 in the text when formulating your answer.)
A: _____

Q: What were the relative humidity, absolute humidity, and wet-bulb temperature of the air at 40°C?

A: Locate the point corresponding to the 40°C air on the slice of the psychrometric chart below.

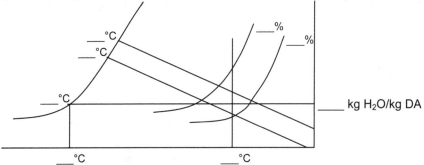

Read and record the following information from the chart:

Relative humidity ≈ ___, Absolute humidity ≈ _____ $\dfrac{\text{kg H}_2\text{O}}{\text{kg dry air}}$, and $T_{\text{wet bulb}}$ ≈ ____°C.

To keep the diagram from getting too cluttered we have not shown the humid volume lines (they slant to the left and are much closer to vertical than the constant wet bulb temperature lines shown on the diagram). Our point is between the lines corresponding to 0.90 m³/kg DA and 0.95 m³/kg DA. Interpolating, we estimate V_h ≈ 0.908 m³/kg DA.

(b) Calculate the mass of the water in the flask.

Solution

This is a good time to draw a flowchart of the process that summarizes what we know about the process. Since everything on the chart is expressed as a ratio to kilograms of dry air, we will label and calculate the mass of dry air in the humid air—once we know that we can determine everything else we need. Neglecting the mass of the mist that formed on the glass when the air reached 20°C, we also know the final point of the process (20°C, saturated, same absolute humidity as the initial air).

Q(kJ)

m_{DA} (kg dry air)
0.0148 kg H₂O(v)/kg DA
2.00 L, 40°C, 1 atm
0.908 m³/kg DA

m_{DA} (kg dry air)
0.0148 kg H₂O(v)/kg DA
20°C, saturated

Mass of dry air

$$m_{DA} = \dfrac{\_\_\_ \text{L}}{} \left| \dfrac{1\ \text{m}^3}{10^3\ \text{L}} \right| \dfrac{1\ \text{kg DA}}{\_\_\_\ \text{m}^3} = _____ \text{kg DA} \qquad (8.72\text{-}3)$$

Mass of water vapor

$$m_{\text{H}_2\text{O}} = \dfrac{\_\_\_\_\ \text{kg DA}}{} \left| \dfrac{\_\_\_\_\ \text{kg H}_2\text{O(v)}}{\text{kg DA}} \right. = _____ \text{kg H}_2\text{O(v)} \qquad (8.72\text{-}4)$$

(c) Calculate the enthalpy change in joules undergone by the air in going from 4°C to 20°C.

Strategy

Think about what you would have to go through to solve Part (b) and this part of the problem without the psychrometric chart. (If you're not sure what you would have to go through, look back at the solution of Problem 8.46.) You would carry out gas law calculations to determine molar quantities, use Raoult's law to determine the water content of the air, set up an enthalpy table, and determine the enthalpies of the water vapor and the air at the initial and final conditions using enthalpy tables or by integrating tabulated heat capacities. Then, you would evaluate $\Delta H = \sum_{out} n_i \hat{H}_i - \sum_{in} n_i \hat{H}_i$. An hour or two later, you'd have your answers. The answers can instead be obtained in minutes using the psychrometric chart. True, the latter values are less precise; however, in practical humidification engineering work, reasonable estimates are often all that is required. In such circumstances, the psychrometric chart can be an invaluable time-saver.

Solution

To find the enthalpies of the humid air at its initial and final conditions, we first use the scale at the upper left of the psychrometric chart to determine $\hat{H}$ at saturation (that is, the specific enthalpy the air would have if we sprayed water into it adiabatically until it became saturated), and then use the *enthalpy deviation curves* on the chart to correct for the fact that the air is not saturated.

(8.72-5)

Q: What is $\hat{H}$ (kJ/kg) at saturation for the air at 40°C? (Locate the point on the chart again and follow the adiabatic saturation line through the point—the same as the constant wet-bulb temperature line—toward the saturation curve. Extrapolate the line past the saturation curve to the "Enthalpy at Saturation" scale and read the value.
A: $\hat{H}_{sat} \approx$ _____ kJ/kg DA
Q: What is $\hat{H}$ for the air at its actual condition? (Estimate the enthalpy deviation for the point on the chart, interpolating between the nearly vertical convex-left curves, and add the value you get to $\hat{H}_{sat}$. The point is between the enthalpy deviation curves corresponding to –0.6 kJ/kg DA and –0.8 kJ/kg DA.)
A: $\hat{H}_{dev} \approx -$_____ $\Rightarrow \hat{H}_{initial} =$ _____ – _____ = _____ kJ/kg DA
Q: What is $\hat{H}$ for the air at its final condition (saturated at 20°C)?
A: $\hat{H}_{sat} \approx$ _____ kJ/kg DA, $\hat{H}_{dev} =$ _____ kJ/kg DA $\Rightarrow \hat{H}_{final} \approx$ _____ kJ/kg DA

(8.72-6)

Q: What is the enthalpy change in joules for the process?
A:
$\Delta H = \left[m_{DA}(\text{kg DA}) \right] \times \left[\Delta \hat{H} \left(\dfrac{\text{kJ}}{\text{kg DA}} \right) \right] = $ _____ $\text{kg DA} \left| \dfrac{(___ - ___)\text{kJ}}{\text{kg DA}} \right| \dfrac{10^3 \text{ J}}{\text{kJ}} = $ _____ J

(d) Write an energy balance for this closed-system process, taking the wet air in the flask as the system, and use it to calculate the heat in joules that must be transferred from the air to accomplish the cooling. (Assume ideal gas behavior, so that $\hat{H} = \hat{U} + P\hat{V} = \hat{U} + RT$.)

Solution

The conversion between H and U involves molar units, so we need to convert from the mass basis we have been using.

(8.72-7)

$$n = \frac{\underline{\hspace{1cm}} \text{ kg dry air}}{} \left| \frac{10^3 \text{ g}}{1 \text{ kg}} \right| \frac{1 \text{ mol}}{\underline{\hspace{0.3cm}} \text{ g}} + \frac{\underline{\hspace{1cm}} \text{ kg H}_2\text{O(v)}}{} \left| \frac{10^3 \text{ g}}{1 \text{ kg}} \right| \frac{1 \text{ mol}}{\underline{\hspace{0.3cm}} \text{ g}} = 0.0777 \text{ mol}$$

$$Q = \Delta U = n\Delta \hat{U} = n\left(\Delta \hat{H} - R\Delta T \right) = \Delta H - nR\Delta T$$

$$= -\underline{\hspace{0.5cm}} \text{ J} - \frac{0.0777 \text{ mol}}{} \left| \frac{\underline{\hspace{1cm}} \text{ J}}{\text{mol} \cdot \text{K}} \right| \frac{\left(\underline{\hspace{0.3cm}} - \underline{\hspace{0.3cm}} \right) °\text{C}}{} \left| \frac{1 \text{ K}}{1°\text{C}} \right| = \underline{\hspace{0.5cm}} \text{ J}$$

PROBLEM 8.86

An 8-molar hydrochloric acid solution (SG = 1.12, $C_p = 2.76$ J/g·°C) is produced by absorbing hydrogen chloride (HCl(g)) in water in a continuous process. Liquid water enters the absorber at 25°C and gaseous HCl is fed at 20°C and 790 torr (absolute). Essentially all of the HCl fed to the column is absorbed. Take one liter of product solution as a basis of calculation.

(a) Estimate the volume (liters) of HCl that must be fed to the absorber.

(b) Estimate the heat (kJ) that must be transferred from the absorber if the product solution is to emerge at 40°C.

(c) Estimate the final solution temperature if the absorber operates adiabatically.

Strategy

When writing energy balances on dissoluton problems involving HCl, NaOH, and H_2SO_4 as solutes and water as the solvent, the heats of solution in Table B.11 will be used, taking the pure solute and liquid water at 25°C as references. The specific enthalpy of a feed or product solution will be determined using a process path that forms the solution at 25°C ($\Delta\hat{H}$ = the tabulated standard heat of solution) and then brings it to the process stream temperature ($\Delta\hat{H}$ = the integral of the solution heat capacity from 20°C to the stream temperature).

(a) Estimate the volume (liters) of HCl that must be fed to the absorber.

Solution

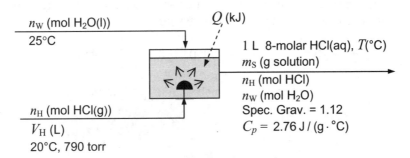

There is enough information given in the product stream specifications to allow the calculation of m_S, n_H, and n_W. Once n_H is known, V_H (L HCl(g)) can be calculated. In the box below, outline how you would do these calculations. (The first step is given. Use as many lines as you need but don't assume that there must be as many steps as there are lines.)

(8.86-1)

- Calculate m_S (g solution) from the volume (1 liter) and specific gravity

- _____

- _____

- _____

- _____

- _____

- _____

- _____

- _____

Now, let's do the calculations.

(8.86-2)

$$m_S = \frac{1 \text{ L solution}}{} \left| \frac{\_\_\_ \text{ cm}^3}{\text{L}} \right| \frac{_____ \text{ g}}{\text{cm}^3} = _____ \text{ g solution}$$

$$n_H = \frac{1 \text{ L solution}}{} \left| \frac{\_\_\_ \text{ mol HCl}}{\text{L}} \right| = \_\_\_ \text{ mol HCl} \Rightarrow m_H = \frac{\_\_\_ \text{ mol HCl}}{} \left| \frac{_____ \text{ g HCl}}{\text{mol}} \right| = _____ \text{ g HCl}$$

$$m_W = (\_\_\_ - \_\_\_) \text{ g H}_2\text{O} = \_\_\_\_\_ \text{ g H}_2\text{O} \Rightarrow n_W = \frac{\_\_\_\_\_ \text{ g H}_2\text{O}}{} \left| \frac{1 \text{ mol H}_2\text{O}}{_____ \text{ g}} \right| = \_\_\_\_\_ \text{ mol H}_2\text{O}$$

$$V_H = \frac{n_H RT}{P} = \frac{8.00 \text{ mol}}{} \left| \frac{\_\_\_\_ \text{ K}}{\_\_\_\_\_ \text{ torr}} \right| \frac{_____ \text{ L} \cdot \text{torr}}{\text{mol} \cdot \text{K}} = _____ \text{ L HCl(g)}$$

(b) Estimate the heat (kJ) that must be transferred from the absorber if the product solution is to emerge at 40°C.

Solution

Setting up the enthalpy table is a little more complex than usual when dealing with mixing and solution problems, for two reasons. First, the heats of solution are expressed per mol of the solute and not of the entire solution, and second, the enthalpy change associated with heating a solution cannot be calculated as the sum of the enthalpy changes of the pure solute and solvent.

References: HCl (g, 25°C, 1 atm), H_2O (l, 25°C, 1 atm)

	n_{in} (mol)	$\hat{H}_{in}$ (kJ/mol)	n_{out} (mol)	$\hat{H}_{out}$ (kJ/mol)
HCl (g)	8.00	$\hat{H}_1$		
H_2O (l)	45.9	0		
HCl (aq)			8.00	$\hat{H}_2$

Enthalpy of feed gas

$$\hat{H}_1 = \int_{\_\_ °C}^{\_\_ °C} (C_p)_{HCl(g)} \, dT = -_____ \frac{\text{kJ}}{\text{mol}}$$

(8.86-3)

The value may be obtained directly from the Physical Property Database of *Interactive Chemical Process Principles* (try it) or by integrating the heat capacity formula in Table B.2.

Enthalpy of product

The process path we will follow to calculate $\hat{H}_2$ (kJ/mol HCl) is:

$$HCl(g, 25°C) + H_2O(l, 25°C) \rightarrow HCl(aq, 25°C) \rightarrow HCl(aq, 40°C)$$

The specific enthalpy change ($\Delta\hat{H}$) of the first process is by definition the standard heat of solution of HCl(g) given in Table B.11. The second specific enthalpy change is obtained by integrating the given solution heat capacity from 25°C to 40°C, multiplying by the mass of the solution to get the total energy change, and then dividing by the moles of HCl to get $\Delta\hat{H}$ in the desired units.

$$r = \frac{46.0 \text{ mol } H_2O}{8 \text{ mol HCl}} = 5.75 \frac{\text{mol } H_2O}{\text{mol HCl}} \xrightarrow{\text{Table B.11}} \Delta \hat{H}_s = -64.87 \frac{\text{kJ}}{\text{mol HCl}}$$

$$\Rightarrow \hat{H}_2 \left(\frac{\text{kJ}}{\text{mol HCl}} \right) = \Delta \hat{H}_s \left(\frac{\text{kJ}}{\text{mol HCl}} \right) + \frac{m_s (\text{g solution}) \int_{25°C}^{40°C} \left[C_p \left(\frac{\text{J}}{\text{g solution} \cdot ° \text{ C}} \right) \right] dT \times \frac{1 \text{ kJ}}{10^3 \text{ J}}}{n_H (\text{mol HCl})}$$

(8.86-4)

$$= -64.87 \frac{\text{kJ}}{\text{mol HCl}} + \frac{(\underline{\quad} \text{ g solution}) \left[\int_{25°C}^{40°C} (\underline{\quad} \, dT) \right] \left(\frac{\text{J}}{\text{g solution}} \right) \frac{1 \text{ kJ}}{10^3 \text{ J}}}{\underline{\quad} \text{ mol HCl}}$$

$$= -64.87 \frac{\text{kJ}}{\text{mol HCl}} + \frac{\underline{\quad} \text{kJ}}{\underline{\quad} \text{mol HCl}} = \underline{\quad} \text{ kJ/mol HCl}$$

The filled-in table is

References: HCl (g, 25°C, 1 atm), H_2O (l, 25°C, 1 atm)

	n_{in} (mol)	$\hat{H}_{in}$ (kJ/mol)	n_{out} (mol HCl)	$\hat{H}_{out}$ (kJ/mol HCl)
HCl (g)	8.00	_____		
H₂O (l)	45.9	0		
HCl (aq)			8.00	_____

and the energy balance is

$$Q = \Delta H = \sum_{\text{out}} n_i \hat{H}_i - \sum_{\text{in}} n_i \hat{H}_i$$

$$= \left(\underline{\quad} \text{ mol HCl} \right) \left(\underline{\qquad} \frac{\text{kJ}}{\text{mol HCl}} \right) - \left(\underline{\quad} \text{ mol HCl} \right) \left(\underline{\qquad} \frac{\text{kJ}}{\text{mol HCl}} \right)$$

(8.86-5)

$$= \underline{\qquad} \text{ kJ}$$

(8.86-6)

Q: Consider the energy of solution that would have been released as heat in the process we have just analyzed if the reactants and products had both been at 25°C (= 519 kJ). Where did the energy go as the process actually ran? (Your answer should involve three different processes along with how much energy went into each one. We'll give you the first one.)

A:

(1) 1.16 kJ was used to raise the HCl(g) fed from 20°C to 25°C

(2) _____ kJ _____

(3) $\dfrac{\underline{\quad} \text{ kJ}}{519 \text{ kJ}}$ _____

(c) Estimate the final solution temperature if the absorber operates adiabatically.

Solution

We could use a shortcut to solve Part (c). The same 519 kJ would be released if the process took place entirely at 25°C, and the same 1.2 kJ goes to raise the temperature of the feed gas from 20°C to 25°C. But, in an adiabatic process, none of the energy is transferred away as heat, so that the entire remaining 517.8 kJ must end up as energy used to heat the product solution. We could therefore integrate the heat capacity of the solution from 25°C to T_{ad} (left as a variable), multiply by the mass of the solution, equate the result to 517.8 kJ, and solve for T_{ad}. Instead, for illustrative purposes we will go through the solution procedure as though we hadn't already done the work of Part (b) (except that we'll copy the specific enthalpy of the feed gas at 20°C).

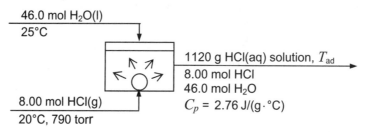

46.0 mol H$_2$O(l)
25°C

1120 g HCl(aq) solution, T_{ad}
8.00 mol HCl
46.0 mol H$_2$O
C_p = 2.76 J/(g·°C)

8.00 mol HCl(g)
20°C, 790 torr

The enthalpy table is filled in with the enthalpy of the final HCl solution ($\hat{H}_2$) as the only unknown (since we copied the enthalpy of the HCl gas ($\hat{H}_1$) from the previous table.)

References: HCl (g, 25°C, 1 atm), H$_2$O (l, 25°C, 1 atm)

	n_{in} (mol)	$\hat{H}_{in}$ (kJ/mol)	n_{out} (mol HCl)	$\hat{H}_{out}$ (kJ/mol HCl)
HCl (g)	8.00	-0.15		
H$_2$O (l)	45.9	0		
HCl (aq)			8.00	$\hat{H}_2$

We determine $\hat{H}_2$ and then solve the energy balance to find T_{ad}.

$$\hat{H}_2\left(\frac{kJ}{mol\ HCl}\right) = \Delta\hat{H}_s\left(\frac{kJ}{mol\ HCl}\right) + \frac{m_s\ (g\ solution)\int_{25°C}^{T_{ad}}\left[C_p\left(\frac{J}{g\ solution\cdot°C}\right)\right]dT\times\frac{1\ kJ}{10^3\ J}}{n_H\ (mol\ HCl)}$$

(8.86-7)

$$= \underline{\hspace{4cm}} = 0.3864T_{ad} - 74.52$$

Energy balance

$$Q = \Delta H = \sum_{out} n_i\hat{H}_i - \sum_{in} n_i\hat{H}_i = 0$$

$$\Rightarrow (8.00\ mol\ HCl)\left[(0.3864T_{ad} - 74.52)\frac{kJ}{mol\ HCl}\right] - \underline{\hspace{2cm}} = 0 \qquad \textbf{(8.86-8)}$$

$$\Rightarrow T_{ad} = \underline{\hspace{1cm}} °C$$

Chapter 9
Balances on Reactive Processes

Name: _____
Date: _____

In Chapter 8, we determined internal energy and enthalpy changes for a variety of processes that did not involve chemical reactions. In this chapter, we tackle similar problems for processes that incorporate exothermic (heat-producing) and endothermic (heat-consuming) chemical reactions.

There are several differences between the procedures for nonreactive and reactive processes, both involving choices that must be made with the reactive ones. With nonreactive processes, all material balances are on the molecular species involved in the process, and all references for enthalpy determinations are the same species at specified conditions (phase, temperature, and pressure). With reactive processes, (1) material balance calculations may involve balances on molecular species (generally inadvisable for reactive substances), atomic constituents of reactive substances, and/or extents of reaction, and (2) references for internal energies and enthalpies of reactive substances may be either the substances themselves (in which case heats of reaction must be included in the energy balance) or their atomic constituents (heats of reaction are not included). Atomic species balances are generally more convenient when more than one reaction is involved in a process. In the problems we work out for this chapter, we will illustrate all of the choices.

PROBLEM 9.2

The standard heat of reaction for the combustion of liquid n-nonane to form CO_2 and liquid water at 25°C and 1 atm is $\Delta \hat{H}_r^\circ = -6124$ kJ/mol.

(a) Briefly explain what that means. Your explanation may take the form "When [____] (specify quantities of reactant species and their physical states) react to form [____] (quantities of product species and their physical state), the change in enthalpy is [____]."

Solution

$$C_9H_{20}(l) + 14O_2(g) \longrightarrow 9CO_2(g) + 10H_2O(l)$$

$$\Delta \hat{H}_r^\circ = -6124 \text{ kJ/mol}$$

(9.2-1)

(b) Is the reaction exothermic or endothermic at 25°C? Would you have to heat or cool the reactor to keep the temperature constant? What would the temperature do if the reactor ran adiabatically? What can you infer about the energy required to break the molecular bonds of the reactants and that released when the product bonds form?

Solution

(9.2-2)

(c) If 25.0 mol/s of liquid nonane are consumed and the reactants and products are all at 25°C, estimate the required rate of heat input or output (state which) in kilowatts, assuming that $Q = \Delta H$ for the process. What have you also assumed about the reactor pressure in your calculation? (You don't have to assume that it equals 1 atm.)

Solution

$$\dot{Q} = \Delta \dot{H} = \frac{\dot{n}_{C_9H_{20}} \Delta \hat{H}_r^\circ}{\nu_{C_9H_{20}}} = \underline{\qquad} \frac{\text{mol } C_9H_{20}}{\text{s}} \left| \frac{\underline{\qquad}\ \text{kJ}}{1\ \text{mol } C_9H_{20}} \right| \frac{1\ \text{kW}}{1\ \text{kJ/s}} \qquad (9.2\text{-}3)$$

$$= \underline{\qquad\qquad} \quad (\text{Heat} \underline{\qquad\qquad})$$

(d) We assumed _____ (9.2-4)

(e) The standard heat of combustion of n-nonane vapor is $\Delta \hat{H}_r^\circ = -6171$ kJ/mol. What is the physical significance of the 47 kJ/mol difference between this heat of combustion and the one given in Part (a)?

Solution

(9.2-5)

```
┌─────────────────────────────────────────────────────┐
│                                                       │
│                                                       │
└─────────────────────────────────────────────────────┘
```

(f) The value of $\Delta \hat{H}_r^\circ$ given in Part (d) applies to n-nonane vapor at 25°C and 1 atm, and yet the normal boiling point of n-nonane is 150.6°C. Can n-nonane exist as a vapor at 25°C and a total pressure of 1 atm? Explain your answer.

Solution

(9.2-6)

```
┌─────────────────────────────────────────────────────┐
│                                                       │
│                                                       │
│                                                       │
│                                                       │
│                                                       │
└─────────────────────────────────────────────────────┘
```

PROBLEM 9.10

The standard heat of combustion ($\Delta\hat{H}_c^\circ$) of liquid 2,3,3-trimethyl pentane (C_8H_{18}) is reported in a table of physical properties to be -4850 kJ/mol. A footnote indicates that the reference temperature for the reported value is 25°C and the presumed combustion products are $CO_2(g)$ and $H_2O(g)$.

(a) In your own words, briefly explain what all that means. (We'll give you a start.)

Solution

$$C_8H_{18}(l) + \frac{25}{2}O_2(g) \longrightarrow 8CO_2(g) + 9H_2O(g) \qquad \Delta\hat{H}_r^\circ = -4850 \text{ kJ/mol}$$

(9.10-1)

When one mol of liquid 2,3,3-trimethyl pentane and 12.5 mol of oxygen at 25°C and 1 atm react to form _____

(b) There is some question about the accuracy of the reported value, and you have been asked to determine the heat of combustion experimentally. You burn 2.010 grams of the hydrocarbon with pure oxygen in a constant-volume calorimeter[9-1] and find that the net heat released when the reactants and products [$CO_2(g)$ and $H_2O(g)$] are all at 25°C is sufficient to raise the temperature of 1.00 kg of liquid water by 21.34°C. Write an energy balance to show that the heat released in the calorimeter equals $n_{C_8H_{18}}\Delta\hat{U}_c^\circ$, and calculate $\Delta\hat{U}_c^\circ$ in kJ/mol. Then calculate $\Delta\hat{H}_c^\circ$. (See Example 9.1-2 in the text.) By what percentage does the tabulated value differ from the measured one?

Strategy

Since we can calculate $n_{C_8H_{18}}$ from the given mass of the fuel and we can calculate the energy released by the reaction (Q) from the information about the temperature rise of the water, we will be able to solve the energy balance equation for $\Delta\hat{U}_c^\circ$.

Solution

Energy balance on water

Whether the heating of the water takes place at constant pressure or constant volume makes no real difference, because for a liquid being heated, $C_p \approx C_v$ and ΔH (which equals Q for a constant pressure process) and ΔU (which equals Q if the process takes place at constant volume) are almost identical.

Table B.2 $\Rightarrow (C_p)_{H_2O(l)} = $ _____ kJ/mol·°C   (9.10-2)

$$Q = \Delta H = m_{H_2O}\int_{T_1}^{T_1+21.34°C} (C_p)_{H_2O(l)} dT = m_{H_2O}(C_p)_{H_2O(l)}(\underline{\hspace{2cm}})$$

(9.10-3)

$$= \frac{1.00 \text{ kg}}{} \left| \frac{1 \text{ mol}}{\underline{\hspace{1cm}} \text{ kg}} \right| \frac{\underline{\hspace{1cm}} \text{ kJ}}{\text{mol·°C}} \left| \frac{\underline{\hspace{1cm}} °C}{} \right| = \underline{\hspace{1cm}} \text{ kJ}$$

[9-1] A calorimeter is a *very well-insulated* reaction vessel. A reaction is carried out inside it, the temperature change from the beginning to the end of the reaction is measured, and the net energy released or absorbed by the reaction is determined from the temperature change and the known amounts and heat capacities of the reactants and products.

Energy balance on calorimeter (Assume constant volume, so that $W = 0$.)

$$Q = \Delta U = n_{C_8H_{18} \text{ consumed}} \frac{\Delta \hat{U}_c^\circ}{\nu_{C_8H_{18}}}$$

$$\Rightarrow \underline{\hspace{1cm}} \text{ kJ} = \frac{\underline{\hspace{1cm}} \text{ g } C_8H_{18} \text{ consumed}}{} \left| \frac{1 \text{ mol } C_8H_{18}}{114.2 \text{ g}} \right| \frac{\Delta \hat{U}_c^\circ \text{ (kJ)}}{\underline{\hspace{0.5cm}} \text{ mol } C_8H_{18} \text{ consumed}}$$

(9.10-4)

$$\Rightarrow \Delta \hat{U}_c^\circ = \underline{\hspace{1cm}} \text{ kJ/mol}$$

Eq. 9.1-5 (text) $\Rightarrow \Delta \hat{H}_c^\circ = \Delta \hat{U}_c^\circ + RT \left[\sum_{\substack{\text{gaseous} \\ \text{products}}} \nu_i - \sum_{\substack{\text{gaseous} \\ \text{reactants}}} \nu_i \right]$

$$\Rightarrow \Delta \hat{H}_c^\circ = \underline{\hspace{1cm}} \text{ kJ/mol} + \frac{\underline{\hspace{1cm}} \text{ J}}{\text{mol} \cdot \text{K}} \left| \frac{1 \text{ kJ}}{10^3 \text{ J}} \right| \underline{\hspace{0.5cm}} \text{ K} \left| \left(\underline{\hspace{0.3cm}} + \underline{\hspace{0.3cm}} - \underline{\hspace{0.3cm}} \right) \right.$$

(9.10-5)

$$= \underline{\hspace{1cm}} \text{ kJ/mol}$$

$$\% \text{ difference} = \frac{(-4850 \text{ kJ / mol}) - (-5068 \text{ kJ / mol})}{\left| -5068 \text{ kJ/mol} \right|} \times 100\% = 4.3\%$$

Taking the measured value to be correct, the error in the reported value is positive (since it is less negative than the measured value).

(c) Use the result of Part (b) to estimate $\Delta \hat{H}_f^\circ$ for 2,3,3-trimethylpentane. Why would the heat of formation of 2,3,3-trimethylpentane probably be determined this way rather than directly from the formation reaction?

Solution

From Eq. (9.3-1) on p. 447 of the text and Table B.1 [for heats of formation of CO_2 and $H_2O(v)$],

$$\Delta \hat{H}_c^\circ = 8 \left(\Delta \hat{H}_f^\circ \right)_{CO_2(g)} + 9 \left(\Delta \hat{H}_f^\circ \right)_{H_2O(v)} - \left(\Delta \hat{H}_f^\circ \right)_{C_8H_{18}(l)}$$

$$\Rightarrow \left(\Delta \hat{H}_f^\circ \right)_{C_8H_{18}(l)} = \left[8 \left(\underline{\hspace{1cm}} \right) + 9 \left(\underline{\hspace{1cm}} \right) - \left(\underline{\hspace{1cm}} \right) \right] \text{kJ/mol}$$

(9.10-6)

$$= \underline{\hspace{1.5cm}} \text{ kJ/mol}$$

(9.10-7)

The reason we don't measure the heat of formation directly is that _____

PROBLEM 9.16

Sulfur dioxide is oxidized to sulfur trioxide in a small pilot plant reactor. SO_2 and 100% excess air are fed to the reactor at 450°C. The reaction proceeds to a 65% SO_2 conversion, and the products emerge from the reactor at 550°C. The production rate of SO_3 is 1.00×10^2 kg/h. The reactor is surrounded by a water jacket into which water at 25°C is fed.

(a) Taking a production rate of 100 kg SO_3/s as a basis of calculation, calculate the flow rates (standard cubic meters per second) of the SO_2 and air feed streams and the extent of reaction, $\dot{\xi}$ (mol/s).

(b) Calculate the standard heat of the SO_2 oxidation reaction, $\Delta\hat{H}_r^\circ$ (kJ/mol). Then, taking molecular species at 25°C as references, prepare and fill in an inlet-outlet enthalpy table, and write an energy balance to calculate the heat (kW) that must be transferred from the reactor to the cooling water.

(c) Calculate the minimum flow rate of the cooling water if its temperature rise is to be kept below 15°C.

(d) Briefly state what would have been different in your calculations and results if you had taken elemental species as references in Part (b).

Strategy

We will convert the given production rate to a molar basis, draw and label the flowchart, and do the required material balance calculations to solve Part (a). We will then do an energy balance on the reactor (excluding the water jacket) to calculate the required rate of heat transfer to solve Part (b), assume that all of that heat goes into the water jacket and none is lost from the jacket to the surroundings, and do an energy balance on the cooling water to solve Part (c).

Solution

$$SO_2(g) + \frac{1}{2}O_2(g) \longrightarrow SO_3(g)$$

Basis

$$\frac{100 \text{ kg } SO_3}{\text{min}} \left| \frac{10^3 \text{ mol } SO_3}{80.07 \text{ kg } SO_3} \right| \frac{1 \text{ min}}{60 \text{ s}} = 20.81 \text{ mol } SO_3/\text{s produced}$$

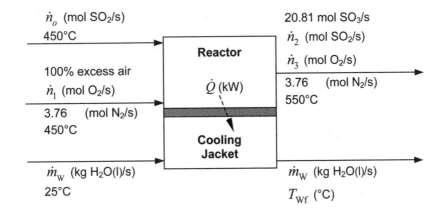

> **Q:** No matter which system we use for material balances (molecular species balances, atomic species balances, or extents of reaction), we can't count a nitrogen balance. Why not?
>
> **A:** _____

We'll use the extent of reaction method to determine the unknown flow rates and carry out the degree-of-freedom analysis accordingly.

(9.16-2)

DEGREE-OF-FREEDOM ANALYSIS ON REACTOR (EXCLUDING JACKET)		
UNKNOWNS AND INFORMATION		**JUSTIFICATION/CONCLUSION**
+ 5 unknowns	$\dot{n}_0 - \dot{n}_3, \dot{\xi}, \dot{Q}$	
+ 1 reaction		Must add when using extent of reaction
− 3 extent of rxn. eqns.		_____
− 1 _____		_____
− 1 _____		_____
− 1 _____		_____
0 DOF		All unknowns can be determined

(a) Taking a production rate of 100 kg SO_3/s as a basis of calculation, calculate the feed rates (standard cubic meters per second) of the SO_2 and air feed streams and the extent of reaction, $\dot{\xi}$ (mol/s).

Solution

Rather than worrying about the order in which to write the equations, let's just write them in any order and let E-Z Solve do the algebra.

100% excess air

$$\dot{n}_1 = \frac{\dot{n}_0 (\text{mol } SO_2 \text{ fed})}{} \left| \frac{\_\_\_ \text{ mol } O_2 \text{ required}}{1 \text{ mol } SO_2 \text{ fed}} \right. \left| \frac{\_\_\_ \text{ mol } O_2 \text{ fed}}{1 \text{ mol } O_2 \text{ required}} = \_\_\_ \dot{n}_0 \right.$$

(9.16-3)

65% SO_2 conversion

$$\dot{n}_2 = \frac{\dot{n}_0 (\text{mol } SO_2 \text{ fed})}{} \left| \frac{\_\_\_ \text{ mol } SO_2 \text{ unreacted}}{\text{mol } SO_2 \text{ fed}} = \_\_\_ \dot{n}_0 \right.$$

(9.16-4)

Extent of reaction equations (from Eq. (4.6-3) on p. 119 in the text)

$$\left(\dot{n}_i \right)_{\text{out}} = \left(\dot{n}_i \right)_{\text{in}} + \upsilon_i \dot{\xi}$$

where υ_i, the stoichometric coefficient of Species i, is positive for products and negative for reactions. We apply this equation to all reactive species.

SO_2 $\dot{n}_{SO_2} = \dot{n}^o_{SO_2} + \upsilon_{SO_2} \dot{\xi} \implies \dot{n}_2 = \dot{n}_0 - \dot{\xi}$ (9.16-5)

O_2 $\dot{n}_{O_2} = \dot{n}^o_{O_2} + \upsilon_{O_2} \dot{\xi} \implies$ _____   (9.16-6)

SO_3   _____ $\implies$ _____   (9.16-7)

Eqs. **(9.16-3)–(9.16-7)** can be solved for the four flow rates and the extent of reaction. The requested volumetric feed rates of SO_2 and air can then be calculated as

$$\dot{V}_0 \text{ (SCMS)} = \frac{\underline{\qquad} \left(\frac{\text{mol } SO_2}{\text{s}}\right) \left| \underline{\qquad} \left(\frac{\text{L(STP)}}{\text{mol}}\right) \right| \frac{1 \text{ m}^3}{}}{\left| 10^3 \text{ L} \right.} \qquad \textbf{(9.16-8)}$$

$$\dot{V}_1 \text{ (SCMS)} \frac{\underline{\qquad} \left(\frac{\text{mol air}}{\text{s}}\right) \left| \underline{\qquad} \left(\frac{\text{L(STP)}}{\text{mol}}\right) \right| \frac{1 \text{ m}^3}{}}{\left| 10^3 \text{ L} \right.} \qquad \textbf{(9.16-9)}$$

Rather than solving these equations now, we will derive the energy balance equations needed for Parts (b) and (c) and have E-Z Solve calculate all of the flowchart variables at once.

(b) Calculate the standard heat of the SO_2 oxidation reaction, $\Delta \hat{H}_r^\circ$ (kJ/mol). Then, taking molecular species at 25°C as references, complete an inlet-outlet enthalpy table, and write an energy balance to calculate the heat (kW) that must be transferred from the reactor to the cooling water.

Solution

From Eq. (9.3-1) in the text and the stoichiometric equation of the reaction,

$$\Delta \hat{H}_r^\circ = \left(\Delta \hat{H}_f^\circ \right)_{\underline{\qquad}} \quad - \left(\Delta \hat{H}_f^\circ \right)_{\underline{\qquad}} \quad \overset{\text{Table B.1}}{\underset{\downarrow}{=}} \quad (\underline{\qquad}) - (\underline{\qquad}) = \underline{\qquad} \text{ kJ/mol} \qquad \textbf{(9.16-10)}$$

(9.16-11)

References $SO_2(g)$, $O_2(g)$, $N_2(g)$, $SO_3(g)$ @ 25°C, 1 atm

	$\dot{n}_{in}$ (mol/s)	$\hat{H}_{in}$ (kJ/mol)	$\dot{n}_{out}$ (mol/s)	$\hat{H}_{out}$ (kJ/mol)
$SO_2(g)$	$\dot{n}_0$	$\hat{H}_A$	_____	$\hat{H}_D$
$O_2(g)$	_____	$\hat{H}_B$	_____	$\hat{H}_E$
$N_2(g)$	_____	$\hat{H}_C$	_____	$\hat{H}_F$
$SO_3(g)$			_____	$\hat{H}_G$

We will use the Physical Property Database of *Interactive Chemical Process Principles* to determine the enthalpy changes for heating the gases from 25°C to 450°C (for the three species at the inlet) and from 25°C to 550°C (for the four species at the outlet). The results are as follows:

(9.16-12)

Reference: 25°C	In (25°C→450°C)	Out (25°C→550°C)
$SO_2(g)$	$\hat{H}_A$ = 19.62 kJ/mol	$\hat{H}_D$ = _____ kJ/mol
$O_2(g)$	$\hat{H}_B$ = _____ kJ/mol	$\hat{H}_E$ = _____ kJ/mol
$N_2(g)$	$\hat{H}_C$ = _____ kJ/mol	$\hat{H}_F$ = _____ kJ/mol
$SO_3(g)$		$\hat{H}_G$ = _____ kJ/mol

Name: _____

Date: _____

(9.16-13)

Q: Instead of using the Physical Property Database, we could have found the required specific enthalpies by using data in the text. How? (Be specific.)

A: For $SO_2(g)$ and $SO_3(g)$, _____

For O_2 and N_2, _____

The open-system energy balance (Eq. 9.5-1a, p. 451 in the text) is (referring to the enthalpy table)

$$\dot{Q}(kW) = \Delta\dot{H} = \dot{\xi}\Delta\hat{H}_r^\circ + \sum_{out} \dot{n}_i\hat{H}_i - \sum_{in} \dot{n}_i\hat{H}_i$$

$$= \left(-98.28\right)\dot{\xi} + \dot{n}_2\hat{H}_D + \dot{n}_3\hat{H}_E + 3.76\dot{n}_1\hat{H}_F + 20.81\hat{H}_G \qquad (9.16\text{-}14)$$

$$- \dot{n}_0\hat{H}_A - \dot{n}_1\hat{H}_B - 3.76\dot{n}_1\hat{H}_C$$

This equation, which introduces the unknown $\dot{Q}$ to the mix, can be solved simultaneously with those derived in Part (a) to determine the heat transferred from the reactor. Eqs. **(9.16-3)–(9.16-7)** and Eq. [9.16-1] total six equations. Count the unknowns in all six equations and convince yourself that, so far, we have six independent equations in six unknowns.

(c) Calculate the minimum flow rate of the cooling water if its temperature rise is to be kept below 15°C.

Solution

A flowchart of the water jacket around the reactor appears as follows:

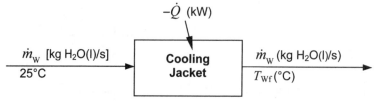

(9.6-15)

Q: Why did we put a minus sign next to $\dot{Q}$ on the flowchart?

A: _____

Q: For a given value of $-\dot{Q}$ [specifically, the one calculated in Part (b)], how would you expect the final cooling water temperature (T_{wf}) to vary with the coolant flow rate ($\dot{m}_w$)?

A: As the flow rate decreases, the final temperature should _____. The minimum flow rate is the one corresponding to the maximum allowable exit temperature of 40°C.

Energy balance on cooling jacket

$$-\dot{Q} = \Delta\dot{H} = \dot{m}_w\left[\underset{\substack{\uparrow \\ \text{Table B.5}}}{\hat{H}_w\left(1, 40°C\right)} - \underset{\substack{\uparrow \\ \text{Table B.5}}}{\hat{H}_w\left(1, 25°C\right)}\right] = \dot{m}_w\left(\frac{kg}{s}\right)\left[\underline{} - \underline{}\right]\frac{kJ}{kg} \qquad (9.16\text{-}16)$$

With this energy balance, we have introduced a seventh equation and a seventh flowchart unknown, $\dot{m}_W$. Eq. **(9.16-16)** completes the set of equations we need to solve for all of the system variables and the variables we were asked to calculate from them. Here is the E-Z Solve program to solve the equations.

E-Z Solve program   **(9.16-17)**

```
//Problem 9.16
//Material balances (Part a of problem)

n1 = n0                        // (Eq. 9.16-3)

n2 = _____        // (Eq. 9.16-4)

n2 = n0 – Xi                   // (Eq. 9.16-5)

n3 = _____        // (Eq. 9.16-6)

Xi = _____        // (Eq. 9.16-7), extent of reaction

V0 = _____        // (Eq. 9.16-8)

V1 = _____        // (Eq. 9.16-9)


//Enthalpies

HA = 19.62 ; HB = _____ ; HC = _____

HD = _____ ; HE = _____ ; HF = _____ ; HG = _____


//Energy balance on reactor (Eq. 9.16-13) (Part b)
Q = –98.28*Xi + n2*HD + n3*HE + 3.76*n1*HF + 20.81*HG – n0*HA – n1*HB – 3.76*n1*HC

//Energy balance on cooling jacket (Part c)

-Q = mw*(_____ – _____)
```

Fill in the blanks, copy the program to E-Z Solve, and run it to get the solutions.

E-Z Solve solution

$\dot{n}_0 =$ _____ mol/s (SO_2 fed) $\dot{n}_1 =$ _____ mol /s (O_2 fed)

$\dot{n}_2 =$ _____ mol SO_2/s (in product gas) $\dot{n}_3 =$ _____ mol SO_2/s (in product gas)   **(9.16-18)**

$\dot{V}_0 =$ _____ SCMS (SO_2 fed) $\dot{V}_1 =$ _____ SCMS (air fed)

$\dot{Q} =$ _____ kW (heat transferred ☐ to ☐ from the reactor)   **(9.16-19)**

$\dot{m}_W =$ _____ kg $H_2O(l)$/s   **(9.16-20)**

                                                                                         (9.16-21)

> **Q:** Briefly state what would have been different in your calculations and results if you had taken elemental species as references in Part (b).
>
> **A:** If elemental (atomic) species had been taken as references, _____

Notes and Calculations

PROBLEM 9.21

Ethanol is produced commercially by the hydration of ethylene.

$$C_2H_4 + H_2O(v) \rightleftharpoons C_2H_5OH(v)$$

Some of the product is converted to diethyl ether in an undesired side reaction.

$$2C_2H_5OH \rightleftharpoons (C_2H_5)_2O(v) + H_2O(v)$$

The combined feed to the reactor contains 53.7 mole% C_2H_4, 37% mole% H_2O and the balance nitrogen and enters the reactor at 310°C. The reactor operates isothermally at 310°C. An ethylene conversion of 5% is achieved, and the yield of ethanol (moles ethanol produced/mole ethylene consumed) is 0.900.

Data for Diethyl Ether

$\Delta\hat{H}_f^\circ = -272.8$ kJ/mol *for the liquid*

$\Delta\hat{H}_v = 26.05$ kJ/mol (assume independent of T)

$(C_p)_{vapor}[kJ/(mol \cdot {}^\circ C)] = 0.08945 + 40.33 \times 10^{-5} T({}^\circ C) - 2.244 \times 10^{-7} T^2$

(a) Calculate the reactor heating or cooling requirement in kJ/mol feed.

Solution

Since there are multiple reactions, it will be convenient to use atomic species balances and to take the elements of the reactive species in their naturally occurring states (C, H_2, and O_2) as references for calculating the enthalpies of those species.

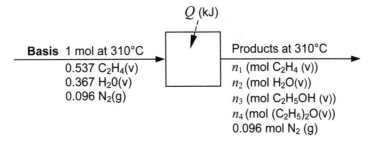

Q (kJ)

Basis 1 mol at 310°C
0.537 C_2H_4(v)
0.367 H_2O(v)
0.096 N_2(g)

Products at 310°C
n_1 (mol C_2H_4 (v))
n_2 (mol H_2O(v))
n_3 (mol C_2H_5OH (v))
n_4 (mol $(C_2H_5)_2O$(v))
0.096 mol N_2 (g)

(9.21-1)

DEGREE-OF-FREEDOM ANALYSIS		
UNKNOWNS AND INFORMATION		**JUSTIFICATION/CONCLUSION**
+ __ unknowns	_____	
− __ reactive atomic species balances	_____	
− 1 C_2H_4 conversion		_____
− 1 C_2H_5OH yield		_____
− 1 energy balance		_____
− 1 DOF		**System is overspecified!**

(9.21-2)

Q: Why didn't we count a nitrogen balance in the DOF analysis?
A: _____

The fact that the number of degrees of freedom is -1 means that we counted one more equation relating system variables than there are variables. This result could mean one of four things:

(1) We forgot to label a system variable on the flowchart. We can easily confirm that this is not the case—all species in the feed and all unconsumed reactants, reaction products, and inert species in the feed are shown and their amounts are labeled on the chart.

(2) The three atomic species balances we have counted are not independent—one of them can be obtained as a linear combination of the others. If this is the case, the four molar quantities can be determined from the specified conversion, the specified yield, and two of the atomic species balances, and the third atomic species balance should close, providing a check on the results.

(3) One more specification was provided than was needed, but it is consistent with the others. This case would look like the previous one—all four amounts could be determined from the first four equations mentioned in Case 2 and the third material balance would provide a check.

(4) At least one of the process specifications is in error. In this case the third material balance would not close.

Strategy

We will solve the conversion and yield equations and two material balances for n_1–n_4, substitute the results into the third material balance, and see if that balance closes. If it does, we can accept the calculated n values and go on to the energy balance.

5% ethylene conversion

$$n_1 = \frac{\underline{\qquad}\ \text{mol } C_2H_4 \ \text{fed} \ \Big| \ \underline{\qquad}\ \text{mol } C_2H_4 \ \text{unreacted}}{\text{mol } C_2H_4 \ \text{fed}}$$

$$= \underline{\qquad}\ \text{mol } C_2H_4 \ \text{unreacted}$$

(9.21-3)

90% ethanol yield

$$n_3 = \frac{\underline{\qquad}\ \text{mol } C_2H_4 \ \text{consumed} \ \Big| \ \underline{\qquad}\ \text{mol } C_2H_5OH \ \text{produced}}{\text{mol } C_2H_4 \ \text{consumed}}$$

$$= \underline{\qquad}\ \text{mol } C_2H_5OH \ \text{produced}$$

(9.21-4)

C balance

$$\frac{\underline{\qquad}\ \text{mol } C_2H_4 \ \Big| \ \underline{\qquad}\ \text{mol C}}{1 \ \text{mol } C_2H_4} = \underline{\ } n_1 + \underline{\ } n_3 + \underline{\ } n_4$$

(9.21-5)

$$\frac{n_1 = \underline{\qquad}}{n_3 = \underline{\qquad}} \rightarrow n_4 = \underline{\qquad}\ \text{mol } (C_2H_5)_2O$$

O balance

$$\frac{\underline{\qquad}\ \text{mol } H_2O \ \Big| \ \underline{\qquad}\ \text{mol O}}{\underline{\ }\ \text{mol } H_2O} = \underline{\qquad\qquad}$$

(9.21-6)

$$\frac{n_3 = \underline{\qquad}}{n_4 = \underline{\qquad}} \rightarrow n_2 = \underline{\qquad}\ \text{mol } H_2O$$

H balance (use to check solutions and consistency of process specifications)

$$(0.537)(4) + (0.367)(2) = \underline{\ } n_1 + \underline{\ } n_2 + \underline{\ } n_3 + \underline{\ } n_4$$

$$\Rightarrow \boxed{2.882} = \underline{\ }(\underline{\qquad}) + \underline{\ }(\underline{\qquad}) + \underline{\ }(\underline{\qquad}) + \underline{\ }(\underline{\qquad}) = \boxed{2.883}$$

(9.21-7)

The H balance closes (the slight difference between the left-hand side and the right-hand side is just round-off error), so that the process specifications are consistent.

(9.21-8)

> **Q:** Are the three atomic species balance equations (on C, O, and H) independent? (*Hint:* No.) Prove your answer by demonstrating that one of them can be obtained as a linear combination of the other two.
>
> **A:** It is easy to show that the H balance can be expressed as (___)(C balance) + (___)(O balance). (Show it using Eqs. **9.21-5**, **9.21-6**, and **9.21-7**.) In the degree-of-freedom analysis, we actually had only **two** *independent* atomic species balances that we were allowed to count, so that the correct number of degrees of freedom is actually 0.

Enthalpy table and enthalpies

When there are multiple reactions, it is almost always more convenient to choose the elements of reactive species in their naturally occurring states at 25°C and 1 atm (the condition at which heats of formation are listed in Table B.1) as references. Any convenient state may still be chosen for non-reactive species.

References: C(s), H_2(g), O_2(g) at 25°C, N_2(g) at 310°C				
	n_{in} (mol)	$\hat{H}_{in}$ (kJ/mol)	n_{out} (mol)	$\hat{H}_{out}$ (kJ/mol)
C_2H_4	0.537	$\hat{H}_1$	0.510	$\hat{H}_1$
H_2O	0.367	$\hat{H}_2$	0.3414	$\hat{H}_2$
N_2	0.096	0	0.096	0
C_2H_5OH			0.02417	$\hat{H}_3$
$(C_2H_5)_2O$			1.415×10^{-3}	$\hat{H}_4$

C_2H_4(g, 310°C)

To determine $\hat{H}_1$, we must calculate the enthalpy change for a process in which C(s) and H_2(g) at 25°C (the reference condition) form ethylene at 310°C. To make use of physical property data in the text, we choose as a process path

$$2C(s, 25°C) + 2H_2(g, 25°C) \rightarrow \underline{\hspace{3cm}} \rightarrow C_2H_4(g, 310°C)$$

(9.21-9)

(9.21-10)

> - $\Delta\hat{H}$ for the first step is by definition _____ and equals _____ kJ/mol.
> - $\Delta\hat{H}$ for the second step is by definition _____ and from Table _____ equals +16.41 kJ/mol. (*Note:* The heat capacity formula for ethylene was programmed incorrectly in the Physical Property Database of *Interactive Chemical Process Principles* and so the enthalpy change calculated by the program is incorrect.)
> - It follows that $\hat{H}_1 = (\underline{\hspace{2cm}} + 16.41)$ kJ/mol = _____ kJ/mol

H₂O (g, 310°C)

$$\hat{H}_2 = \underline{\hspace{2cm}} + \int_{25}^{310} \underline{\hspace{3cm}} dT$$

$$\underbrace{\text{Table B.1}}_{\text{Table B.8}} \longrightarrow = \left(\underline{\hspace{2cm}} + \underline{\hspace{2cm}} \right) \text{kJ/mol} = \underline{\hspace{2cm}} \text{kJ/mol}$$

(9.21-11)

C₂H₅OH (g, 310°C)

$$\hat{H}_3 = \underline{\hspace{2cm}} + \int_{25}^{310} \underline{\hspace{3cm}} dT$$

$$\underbrace{\text{Table B.1}}_{\text{Table B.2}} \longrightarrow \left(\underline{\hspace{2cm}} + 24.16 \right) \text{kJ/mol} = \underline{\hspace{2cm}} \text{kJ/mol}$$

(9.21-12)

(*Note:* The heat capacity formula for ethanol was programmed incorrectly in the Physical Property Database of *Interactive Chemical Process Principles* and so the enthalpy change calculated by the program is incorrect.)

(C₂H₅)₂O (g,310°C)
From the physical property data given for diethyl ether (the standard heat of formation of the liquid, the heat of vaporization, and the heat capacity of the vapor), the process path must be

(9.21-13)

> - formation of the liquid from its elements at 25°C, followed by,
> - _____, followed by
> - _____.

$$\Rightarrow \hat{H}_4 = \underline{\hspace{2cm}} + \underline{\hspace{2cm}} + \int_{25}^{310} \underline{\hspace{2cm}} dT$$

$$= \left(\underline{\hspace{1cm}} + \underline{\hspace{1cm}} + \underline{\hspace{1cm}} \right) \text{kJ/mol} = \underline{\hspace{1cm}} \text{kJ/mol}$$

(9.21-14)

Energy balance

$$Q = \Delta H = \sum_{\text{out}} n_i \hat{H}_i - \sum_{\text{in}} n_i \hat{H}_i$$

$$= (0.510 \text{ mol}) \left(68.69 \frac{\text{kJ}}{\text{mol}} \right) + (0.3414)(-233.58) + (0.02417)(-211.15)$$

$$+ (1.45 \times 10^{-3})(-204.2) - (0.537)(68.69) - (0.367)(-233.58)$$

$$= -1.27 \text{ kJ} \Rightarrow \underline{1.27 \text{ kJ transferred from reactor/mol feed}}$$

(b) Why would the reactor be designed to yield such a low conversion of ethylene? What processing step (or steps) would probably follow the reactor in a commercial implementation of this process?

Solution

(9.21-15)

> If the conversion of ethylene is low, _____
>
>
> Following the reactor, _____

PROBLEM 9.33

The synthesis of methanol from carbon monoxide is carried out in a continuous vapor-phase reactor at 5.00 atm absolute. The feed contains CO and H_2 in stoichiometric proportion and enters the reactor at 25°C and 5.00 atm at a rate of 17.1 m³/h. The product stream emerges from the reactor at 127°C. The rate of heat transfer from the reactor is 17.05 kW. Calculate the fractional conversion achieved and the volumetric flow rate (m³/h) of the product stream. (See Example 9.5-4 in the text.)

Strategy

In most of the problems we have worked thus far, we were given enough information to determine all of the material flows and specific enthalpies of the components of all process streams, after which we could write the energy balance to determine the required heat flow. In another class of problems (which includes this one) the heat flow is specified, either by giving its value (as in this problem) or by stating that the process is adiabatic (in which case the heat flow is zero). In such problems, we must solve the energy balance and the other system equations simultaneously for all unknown process variables. We'll set up both the material and energy balance calculations for this problem in terms of the extent of reaction.

Solution

$$CO(g) + 2H_2(g) \longrightarrow CH_3OH(g)$$

$$\Delta \hat{H}_r^\circ = \left(\Delta \hat{H}_f^\circ\right)_{\underline{\quad\quad}} - \left(\Delta \hat{H}_f^\circ\right)_{\underline{\quad\quad}} = (\underline{\quad} - \underline{\quad})kJ/mol = \underline{\quad\quad} \; kJ/mol \qquad \text{(9.33-1)}$$

Basis

$$\frac{17.1 \text{ m}^3}{\text{h}} \left| \frac{10^3 \text{ L}}{1 \text{ m}^3} \right| \frac{\underline{\quad} \text{ K}}{\underline{\quad} \text{ K}} \left| \frac{\underline{\quad} \text{ atm}}{\underline{\quad} \text{ atm}} \right| \frac{1 \text{ mol}}{\underline{\quad} \text{ L(STP)}} = 3497 \text{ mol/h feed} \qquad \text{(9.33-2)}$$

3497 mol/h → [reactor] → $\dot{n}_1$ (mol CO/h)
0.333 mol CO/mol $\dot{n}_2$ (mol H_2/h)
0.667 mol H_2/mol $\dot{n}_3$ (mol CH_3OH(v)/h)
25°C, 5 atm   127°C, 5 atm
$\dot{Q} = -17.05$ kW

(9.33-3)

DEGREE-OF-FREEDOM ANALYSIS		
UNKNOWNS AND INFORMATION		JUSTIFICATION/CONCLUSION
+4 unknowns	_____	(Counting extent of reaction)
−3 extent of reaction equations	_____	
−1 energy balance		_____
0 DOF		Determine all unknowns

Extent of reaction equations [Eq. (4.6-3) in the text: $\left(\dot{n}_i\right)_{out} = \left(\dot{n}_i\right)_{in} + v_i \dot{\xi}$]:

CO $\dot{n}_1 = \left(3497 \dfrac{\text{mol}}{\text{h}}\right)\left(0.3333 \dfrac{\text{mol CO}}{\text{mol}}\right) + (-1)\dot{\xi} = 1166 - \dot{\xi}$

H_2 $\dot{n}_2 =$ _____   (9.33-4)

CH_3OH $\dot{n}_3 =$ _____   (9.33-5)

CO (g,127°C) $\hat{H}_1 = \hat{H}_{CO}(127°C) \overset{\text{ICPP or Table B.8}}{\underset{\downarrow}{=}} 2.987 \text{ kJ/mol}$

H₂ (g,127°C) $\hat{H}_2 = \hat{H}_{H_2}(127°C) \overset{\text{ICPP or Table B.8}}{\underset{\downarrow}{=}} \rule{1.5cm}{0.4pt} \text{ kJ/mol}$

CH₃OH (g,127°C) $\hat{H}_3 = \int_{25}^{127} C_p dT \overset{\text{ICPP or Table B.2}}{\underset{\downarrow}{=}} \rule{1.5cm}{0.4pt} \text{ kJ/mol}$

(9.33-7)

Energy balance (You might find it easy to use E-Z Solve)

$$\dot{Q} = \Delta\dot{H} \overset{\text{Eq. (9.5-1a)}}{\underset{\downarrow}{=}} \dot{\xi}\Delta\hat{H}_r^\circ + \sum_{\text{out}} \dot{n}_i\hat{H}_i - \sum_{\text{in}} \dot{n}_i\hat{H}_i$$

$$\Rightarrow \frac{\rule{1.5cm}{0.4pt} \text{ kJ}}{\text{s}} \left| \frac{3600 \text{ s}}{1 \text{ h}} \right. = \left[\dot{\xi}\left(\frac{\text{mol}}{\text{h}}\right)\right]\left(\frac{\rule{1cm}{0.4pt} \text{ kJ}}{\text{mol}}\right) + \left[(\rule{1cm}{0.4pt})\frac{\text{mol}}{\text{h}}\right]\left(2.987\frac{\text{kJ}}{\text{mol}}\right)$$

$$+ (\rule{1.5cm}{0.4pt})(\rule{1.5cm}{0.4pt}) + (\rule{1.5cm}{0.4pt})(\rule{1.5cm}{0.4pt})$$

(9.33-8)

$$\Rightarrow \dot{\xi} = \rule{1.5cm}{0.4pt} \text{ mol/h}$$

Fractional conversion

Since CO and H₂ are fed in stoichiometric proportion, the fractional conversion of one equals the fractional conversion of the other. We will do the calculation for CO.

$$f_{CO} = \frac{\left[(\dot{n}_{CO})_{\text{in}} - (\dot{n}_{CO})_{\text{out}}\right]}{(\dot{n}_{CO})_{\text{in}}} = \frac{(\rule{1cm}{0.4pt}) \text{ mol}_{\text{in}}/\text{h} - (\rule{1cm}{0.4pt}) \text{ mol}_{\text{out}}/\text{h}}{(\rule{1cm}{0.4pt}) \text{ mol}_{\text{in}}/\text{h}}$$

(9.33-9)

$$= \rule{1.5cm}{0.4pt} \text{ mol reacted/mol fed}$$

Volumetric flow rate of product stream

The total molar flow rate of the product stream is obtained by summing the three individual molar flow rates (using the extent of reaction equations), and that in turn is converted to the volumetric flow rate using the (now) known molar flow rate and the ideal gas equation of state.

$$\dot{n}_{\text{out}} = (\dot{n}_1 + \dot{n}_2 + \dot{n}_3) = (1166 - \dot{\xi}) + \rule{1.5cm}{0.4pt} + \rule{1.5cm}{0.4pt} = 3497 - 2\dot{\xi} = \rule{1.5cm}{0.4pt} \text{ mol/h}$$

(9.33-10)

$$\Rightarrow \dot{V}_{\text{out}} = \frac{\dot{n}_{\text{out}}RT}{P} = \frac{\rule{1cm}{0.4pt} \text{ mol}}{\text{h}} \left| \frac{\rule{1cm}{0.4pt} \text{ K}}{\rule{0.5cm}{0.4pt} \text{ atm}} \right. \left| \frac{\rule{1cm}{0.4pt} \text{ L·atm}}{\text{mol·K}} \right| \frac{1 \text{ m}^3}{10^3 \text{ L}} = \rule{1.5cm}{0.4pt} \text{ m}^3/\text{h}$$

PROBLEM 9.45

A 2.00 mole% sulfuric acid solution is neutralized with a 5.00 mole% sodium hydroxide solution in a continuous reactor. All reactants enter at 25°C. The standard heat of solution of sodium sulfate is −1.17 kJ/mol Na_2SO_4, and the heat capacities of all solutions may be taken to be that of pure liquid water [4.184 kJ/(kg·°C)].

(a) How much heat (kJ/kg acid solution fed) must be transferred to or from the reactor contents (state which it is) if the product solution emerges at 40°C.

Solution

The reaction stoichiometry is:

$$H_2SO_4(aq) + 2NaOH(aq) \longrightarrow Na_2SO_4(aq) + 2H_2O(aq)$$

Basis

1 mol H_2SO_4 fed $\xrightarrow{\text{2.00 mole\% solution}}$ ____ mol acid solution $\Rightarrow$ ____ mol H_2O in acid solution

$\xrightarrow{\text{acid is neutralized}}$ 2 mol NaOH fed $\xrightarrow{\text{5 mole\% solution}}$ ____ mol basic solution   **(9.45-1)**

$\Rightarrow$ ____ mol H_2O in base solution

The reaction products can be determined easily from the feed components and the reaction stoichiometry. Fill in the values on the flowchart. (Don't forget that water is formed in the reaction.)

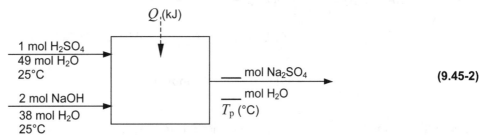

Q_i(kJ)

1 mol H_2SO_4
49 mol H_2O
25°C

2 mol NaOH
38 mol H_2O
25°C

____ mol Na_2SO_4 **(9.45-2)**
____ mol H_2O
T_p (°C)

The only unknown is Q, and so an energy balance on the reactor is all that is needed to determine it. The tricky part of the calculation is keeping track of the units, since heats of solution are given in kJ/mol solute and heat capacities are expressed in kJ/(kg solution·°C). We will choose as references the elemental species that constitute the feed and product stream components [$H_2(g)$, $S(s)$, $O_2(g)$, $Na(s)$] at 25°C and 1 atm as reference states for the enthalpy calculations. We'll determine the solution enthalpies by forming the solutes and water at 25°C from the elements, forming the feed and product solutions at 25°C, and heating the product solution to its final temperature. A block diagram of the procedure is:

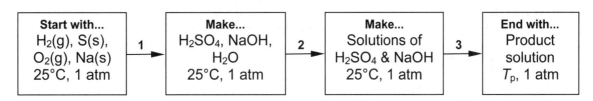

Complete the mass balances for the flowchart and write the molar quantities in the enthalpy table that follows **(9.45-3)**. (Note that you only need moles of solute, since all enthalpies in the table will be expressed as kJ per mole of solute.)

Name: _____

Date: _____

(9.45-3)

References: $H_2(g)$, $S(s)$, $O_2(g)$, $Na(s)$ @ 25°C, 1 atm

	n_{in} (mol solute)	$\hat{H}_{in}$ (kJ/mol solute)	n_{out} (mol solute)	$\hat{H}_{out}$ (kJ/mol solute)
H_2SO_4 (aq)	___	$\hat{H}_1$		
NaOH (aq)	___	$\hat{H}_2$		
Na_2SO_4 (aq)			___	$\hat{H}_3$

Now, let's calculate the specific enthalpies we need for the table. We've combined some of the steps in the block diagram in the calculation below. Identify the steps below in the diagram as you fill in the calculation for the sulfuric acid.

H_2SO_4 (aq, 25°C, r = 49 mol H_2O/mol H_2SO_4)

$$H_2(g) + \underline{\quad} + \underline{\quad} \rightarrow H_2SO_4(l,25°C)$$
$$49H_2(g) + \underline{\quad} O_2(g) \rightarrow 49H_2O(l,25°C)$$
$$\left. \right\} \rightarrow H_2SO_4(aq, 25°C, r = 49) \qquad (9.45\text{-}4)$$

$$\Delta\hat{H} = \hat{H}_1 = \left(\Delta\hat{H}_f^\circ\right)_{H_2SO_4(l)} + 49\left(\Delta\hat{H}_f^\circ\right)_{H_2O(l)} + \Delta\hat{H}_s^\circ\left(r = 49\right)$$

$$= [\underline{\qquad} + 49\left(\Delta\hat{H}_f^\circ\right)_{H_2O(l)} + (\underline{\qquad})] \text{ kJ/mol } H_2SO_4 \qquad (9.45\text{-}5)$$

$$= \left[\underline{\qquad} + 49\left(\Delta\hat{H}_f^\circ\right)_{H_2O(l)}\right] \text{kJ/mol } H_2SO_4$$

(You will find the heats of formation in Table B.1 and the heat of solution in Table B.11.) Enter the expression for $\hat{H}_1$ in the enthalpy table.

Note: The term in Eq. **(9.45-5)** that contains the heat of formation of liquid water $[49\left(\Delta\hat{H}_f^\circ\right)_{H_2O(l)}]$ is numerically much larger than the other terms, and the same will be true for the other two enthalpy expressions we will write. Rather than substituting a value for $[49\left(\Delta\hat{H}_f^\circ\right)_{H_2O(l)}]$ now and possibly incurring round-off error when the enthalpy expressions are substituted into the energy balance, we've left the heat of formation of water as a variable. The terms for water fed as solvent will cancel out, so that only the heat of formation of the two moles of water formed in the reaction will contribute to the final expression for Q.

NaOH (aq, 25°C, r = 19 mol H_2O/mol NaOH)

$$\underline{\qquad\qquad} \rightarrow NaOH(s,25°C)$$
$$\underline{\qquad\qquad} \rightarrow 19H_2O(l,25°C)$$
$$\left. \right\} \rightarrow NaOH(aq, 25°C, r = 19) \qquad (9.45\text{-}6)$$

$$\Delta\hat{H} = \hat{H}_2 = \left(\Delta\hat{H}_f^\circ\right)_{NaOH(s)} + 19\left(\Delta\hat{H}_f^\circ\right)_{H_2O(l)} + \Delta\hat{H}_s^\circ\left(r = 19\right)$$

$$= \left[\underline{\hspace{1cm}} + 19\left(\Delta\hat{H}_f^\circ\right)_{H_2O(l)} + \left(\underline{\hspace{1cm}}\right)\right] \text{kJ/mol NaOH} \qquad \text{(9.45-7)}$$

$$= \left[\underline{\hspace{1cm}} + 19\left(\Delta\hat{H}_f^\circ\right)_{H_2O(l)}\right] \text{kJ/mol NaOH}$$

Na$_2$SO$_4$ (aq, 25°C)

$$\underline{\hspace{3cm}} \rightarrow Na_2SO_4(s,25°C)$$

$$\underline{\hspace{3cm}} \rightarrow 89H_2O(l,25°C)$$

$$\rightarrow Na_2SO_4(aq,25°C) \rightarrow Na_2SO_4(aq,40°C) \quad \text{(9.45-8)}$$

$$m_{soln} = \frac{1 \text{ mol Na}_2SO_4}{} \left|\frac{\underline{\hspace{0.5cm}} \text{ g}}{\text{mol}}\right| \frac{1 \text{ kg}}{10^3 \text{ g}} + \frac{\underline{\hspace{0.5cm}} \text{ mol H}_2O}{} \left|\frac{\underline{\hspace{0.5cm}} \text{ g}}{\text{mol}}\right| \frac{1 \text{ kg}}{10^3 \text{ g}} \qquad \text{(9.45-9)}$$

$$= \underline{\hspace{1cm}} \text{ kg solution}$$

$$\Delta\hat{H} = \hat{H}_3 = \left[\underline{\hspace{1cm}} + 89\left(\underline{\hspace{1cm}}\right) + \underline{\hspace{1cm}}\right]\frac{\text{kJ}}{\text{mol Na}_2SO_4}$$

$$+ \frac{\left[m_{soln}\int_{\underline{\hspace{0.3cm}}}^{\underline{\hspace{0.3cm}}}\left(C_p\right)_{soln} dT\right](\text{kJ})}{1 \text{ mol Na}_2SO_4}$$

$$= \left[\underline{\hspace{1cm}} + 89\left(\Delta\hat{H}_f^\circ\right)_{H_2O(l)} + (-1.17)\right]\frac{\text{kJ}}{\text{mol Na}_2SO_4} \qquad \text{(9.45-10)}$$

$$+ \left(1.746 \text{ kg}\right)\left(4.184 \frac{\text{kJ}}{\text{kg} \cdot °C}\right)(40°C - 25°C)$$

$$= \left[\underline{\hspace{1cm}} + 89\left(\Delta\hat{H}_f^\circ\right)_{H_2O(l)}\right] \text{kJ/mol Na}_2SO_4$$

Energy balance

$$Q = \Delta H = \sum_{out} n_i\hat{H}_i - \sum_{in} n_i\hat{H}_i \qquad \text{[Substitute values from enthalpy table (9.45 - 3)]}$$

$$= (1 \text{ mol Na}_2SO_4)\left[-1276 + 89\left(\Delta\hat{H}_f^\circ\right)_{H_2O(l)}\right]\left(\frac{\text{kJ}}{\text{mol Na}_2SO_4}\right)$$

$$- (\underline{\hspace{0.5cm}} \text{ mol H}_2SO_4)\left[\underline{\hspace{2cm}}\right]\left(\frac{\text{kJ}}{\text{mol H}_2SO_4}\right) \qquad \text{(9.45-11)}$$

$$- (\underline{\hspace{0.5cm}} \text{ mol NaOH})\left[\underline{\hspace{2cm}}\right]\left(\frac{\text{kJ}}{\text{mol NaOH}}\right)$$

$$= 547 + 2\left(\Delta\hat{H}_f^\circ\right)_{H_2O(l)} = \underline{\hspace{1cm}} \text{ kJ (transferred from reactor)}$$

(b) Estimate the product solution temperature if the reactor is adiabatic, neglecting heat transferred between the reactor contents and the reactor wall.

Solution

This calculation is identical to the previous one, except that Q now equals zero and the integral in the calculation of the enthalpy of sodium sulfate goes from 25°C to T_p instead of from 25°C to 40°C. Consequently, we can carry over the expressions for the enthalpies of the feed components and redo the last two calculations as shown below.

Na$_2$SO$_4$ [aq, T_p (°C)]

$$\Delta \hat{H} = \hat{H}_3 = \left[\underline{\hspace{4cm}} \right] + \frac{\left[m_{\text{soln}} \int_{25}^{T_p} \left(C_p \right)_{\text{soln}} dT \right]}{1}$$

$$= [-1384.5 + 89 \left(\Delta \hat{H}_f^\circ \right)_{H_2O(l)} + (-1.17)] + (\underline{\hspace{1cm}})(\underline{\hspace{1cm}})(T_p - 25) \qquad \textbf{(9.45-12)}$$

$$= \left[-1385.7 + 89 \left(\Delta \hat{H}_f^\circ \right)_{H_2O(l)} \right] + \underline{\hspace{1cm}}(T_p - 25)$$

Energy balance

$$Q = 0 = \sum_{\text{out}} n_i \hat{H}_i - \sum_{\text{in}} n_i \hat{H}_i$$

$$= (\underline{\hspace{0.5cm}} \text{ mol Na}_2\text{SO}_4) \left[-1385.7 + 89 \left(\Delta \hat{H}_f^\circ \right)_{H_2O(l)} + \underline{\hspace{1cm}}(T_p - 25) \right] \left(\frac{\text{kJ}}{\text{mol Na}_2\text{SO}_4} \right)$$

$$- (\underline{\hspace{0.5cm}} \text{ mol H}_2\text{SO}_4) \left[-884.6 + 49 \left(\Delta \hat{H}_f^\circ \right)_{H_2O(l)} \right] \left(\frac{\text{kJ}}{\text{mol H}_2\text{SO}_4} \right) \qquad \textbf{(9.45-13)}$$

$$- (\underline{\hspace{0.5cm}} \text{ mol NaOH}) \left[-469.4 + 19 \left(\Delta \hat{H}_f^\circ \right)_{H_2O(l)} \right] \left(\frac{\text{kJ}}{\text{mol NaOH}} \right)$$

$$\Rightarrow T_p = \underline{\hspace{1.5cm}} \text{ °C}$$

PROBLEM 9.55
Methane at 25°C is burned in a boiler furnace with 10.0% excess air preheated to 100°C. Ninety percent of the methane fed is consumed, the product gas contains 10.0 mol CO_2/mol CO, and the combustion products leave the furnace at 400°C.

(a) Calculate the heat transferred from the furnace, $\dot{Q}(kW)$, for a basis of 1000 mol CH_4 fed/s. (The greater the value of $\dot{Q}$, the more steam is produced in the boiler.)

Solution

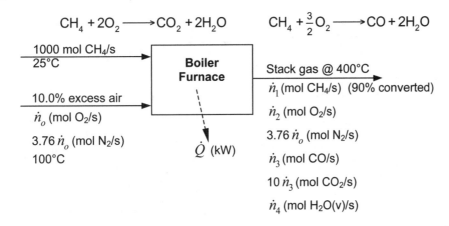

$$CH_4 + 2O_2 \longrightarrow CO_2 + 2H_2O \qquad CH_4 + \frac{3}{2}O_2 \longrightarrow CO + 2H_2O$$

1000 mol CH_4/s
25°C

10.0% excess air
$\dot{n}_o$ (mol O_2/s)

3.76 $\dot{n}_o$ (mol N_2/s)
100°C

Boiler Furnace

$\dot{Q}$ (kW)

Stack gas @ 400°C
$\dot{n}_1$ (mol CH_4/s) (90% converted)
$\dot{n}_2$ (mol O_2/s)
3.76 $\dot{n}_o$ (mol N_2/s)
$\dot{n}_3$ (mol CO/s)
10 $\dot{n}_3$ (mol CO_2/s)
$\dot{n}_4$ (mol $H_2O(v)$/s)

In the degree-of-freedom analysis, don't count the nitrogen balance and the CO_2/CO selectivity when listing relations among the variables.

(9.55-1)

DEGREE-OF-FREEDOM ANALYSIS		
UNKNOWNS AND INFORMATION		JUSTIFICATION/CONCLUSION
+ 6 unknowns	$\dot{n}_o - \dot{n}_4, \dot{Q}$	
− 3 atomic species balances	C,H,O	
− 1 _____		_____
− 1 _____		_____
− 1 _____		_____
0 DOF		Determine all unknowns

(9.55-2)

Q: Why couldn't we count a nitrogen balance or the given CO_2/CO selectivity in the DOF analysis?

A: _____

We will derive the equations and let E-Z Solve do the number crunching, but we'll also show the order in which the equations should be written and solved if the solution is done manually. Circle the variable for which each equation would be solved.

Oxygen feed rate

$$\dot{n}_o = \frac{1000 \text{ mol } CH_4}{s} \left| \frac{\_\_ \text{ mol } O_2 \text{ required}}{\text{mol } CH_4} \right| \frac{\_\_\_ \text{ mol } O_2 \text{ fed}}{\text{mol } O_2 \text{ required}}$$

(9.55-3)

90% methane consumption

$$\dot{n}_1 = (1000 \text{ mol } CH_4/\text{fed}) \left[\underline{\hspace{2cm}} \frac{\text{mol } CH_4 \text{ unconsumed}}{\text{mol } CH_4/\text{fed}} \right]$$

(9.55-4)

C balance

$$\frac{1000 \text{ mol } CH_4}{s} \left| \frac{1 \text{ mol } C}{1 \text{ mol } CH_4} = \dot{n}_1(1) + \dot{n}_3 (\_\_) + \underline{\hspace{2cm}} \right.$$

(9.55-5)

H balance

$$\frac{1000 \text{ mol } CH_4}{s} \left| \frac{\underline{\hspace{1.5cm}}}{\underline{\hspace{1.5cm}}} = \underline{\hspace{2.5cm}} \right.$$

(9.55-6)

O balance

$$\underline{\hspace{2.5cm}} = \underline{\hspace{4cm}}$$

(9.55-7)

Enthalpy table and enthalpies

(9.55-8)

References: C(s), H₂(g), O₂(g), N₂(g) @ 25°C, 1 atm

	$\dot{n}_{in}$ (mol/s)	$\hat{H}_{in}$ (kJ/mol)	$\dot{n}_{out}$ (mol/s)	$\hat{H}_{out}$ (kJ/mol)
CH₄(g)	1000	$\hat{H}_A$	$\dot{n}_1$	$\hat{H}_D$
O₂(g)	___	$\hat{H}_B$	___	$\hat{H}_E$
N₂(g)	___	$\hat{H}_C$	___	$\hat{H}_F$
CO(g)			___	$\hat{H}_G$
CO₂(g)			___	$\hat{H}_H$
H₂O(v)			___	$\hat{H}_I$

$$\hat{H}_A = \left(\Delta \hat{H}_f^\circ \right)_{CH_4} \xrightarrow{\text{Table} \_\_\_} \underline{\hspace{1.5cm}} \text{ kJ/mol}$$

(9.55-9)

$$\hat{H}_B = \hat{H}(O_2, 100°C) \xrightarrow{\text{Table B.8}} \underline{\hspace{1cm}} \text{ kJ/mol}$$

(9.55-10)

$$\hat{H}_C = \hat{H}(N_2, 100°C) \xrightarrow{\text{Table} \_\_\_} \underline{\hspace{1cm}} \text{ kJ/mol}$$

(9.55-11)

$$\hat{H}_D = \left(\Delta \hat{H}_f^\circ \right)_{CH_4} + \int_{25}^{400} \left(C_p \right)_{CH_4} dT \xrightarrow[\text{Phys. Prop. Database}]{\text{Table B.1}} (\underline{\hspace{1cm}} + \underline{\hspace{1cm}}) \text{kJ/mol}$$

(9.55-12)

$$\hat{H}_E = \hat{H}(O_2, 400°C) \xrightarrow{\text{Table} \_\_\_\_} _____ \text{kJ/mol}$$ **(9.55-13)**

$$\hat{H}_F = \hat{H}(N_2, 400°C) \xrightarrow{\text{Table} \_\_\_\_} _____ \text{kJ/mol}$$ **(9.55-14)**

$$\hat{H}_G = _____ \xrightarrow[\text{Table} \_\_\_\_]{\text{Table} \_\_\_\_} (_____ + _____) \text{kJ/mol}$$ **(9.55-15)**

$$\hat{H}_H = _____ \xrightarrow[\text{Table} \_\_\_\_]{\text{Table} \_\_\_\_} _____ \text{kJ/mol}$$ **(9.55-16)**

$$\hat{H}_I = _____ \xrightarrow[\text{Table} \_\_\_\_]{\text{Table} \_\_\_\_} _____ \text{kJ/mol}$$ **(9.55-17)**

Energy balance

Since we are using atomic species as references, there is no $\Delta \hat{H}_r°$ term.

$$\boxed{\dot{Q}} = \Delta \dot{H} = \sum_{\text{out}} \dot{n}_i \hat{H}_i - \sum_{\text{in}} \dot{n}_i \hat{H}_i$$

$$= \dot{n}_1 \hat{H}_D + \dot{n}_2 \hat{H}_E + _____$$ **(9.55-18)**

$$- 1000 \hat{H}_A - _____$$

E-Z Solve program

```
// Problem 9.55
// Molar flow rates (ni in mol/s)
n0 = 1000*2*1.1                    // Eq. (9.55-3)

n1 = _____         // Eq. (9.55-4)

_____             // Eq. (9.55-5)

_____             // Eq. (9.55-6)

_____             // Eq. (9.55-7)


// Enthalpies (Hi in kJ/mol)

HA = _____

HB = _____

HC = _____

HD = _____

HE = _____

HF = _____

HG = _____

HH = _____

HI = _____

// Energy balance (Q in kW)
Q = n1*HD + n2*HE + 3.76*n0*HF + n3*HG + 10*n3*HH + n4*HI - 1000*HA - n0*HB -3.76*n0*HC
```

E-Z Solve solution

$$\dot{Q} = _____ \text{kW}$$ **(9.55-19)**

(b) Would the following changes increase or decrease the rate of steam production? (Assume the fuel feed rate and fractional conversion of methane remain constant.) Briefly explain your answers. **(i)** Increasing the temperature of the inlet air; **(ii)** Increasing the percent excess air for a given stack gas temperature; **(iii)** Increasing the selectivity of CO_2 to CO formation in the furnace; **(iv)** Increasing the stack gas temperature.

(9.55-20)

(i)   If T_{air} increases, the heat transferred ☐ increases ☐ decreases because _____
(ii) If the percentage excess air increases, the heat transferred _____
(iii) If more CO_2 and less CO are produced, the heat transferred _____
(iv) If the stack gas temperature is increased, the heat transferred _____

PROBLEM 9.66

Methane is burned with 25% excess air in a continuous adiabatic reactor. The methane enters the reactor at 25°C and 1.10 atm at a rate of 5.50 liters/s, and the entering air is at 150°C and 1.1 atm. Combustion in the reactor is complete, and the reactor effluent gas emerges at 1.05 atm. Calculate (a) the temperature, and (b) the degrees of superheat of the reactor effluent. (Consider water to be the only condensable species in the effluent.)

Strategy

The material balance calculations are carried out normally. Once the composition of the product gas is known, the dew point can be determined from the known pressure and Raoult's law (Eq. 6.3-3 in the text). Since the product temperature (T_f) is unknown, the enthalpies of the product stream components are determined by integrating their heat capacities from the reference temperature to T_f. The resulting expressions for $\hat{H}_i(T_f)$ are substituted into the energy balance ($\dot{Q} = \Delta \dot{H} = 0$), which is then solved for T_f. After T_f has been determined, the degrees of superheat of the product stream can be calculated. Since the energy balance equation will be a fourth-order polynomial in T_f and we will need a computer to solve it, we will again collect all the system equations and let E-Z Solve handle the simultaneous solution and root-finding required.

Solution

Since we only have one reaction and the heat of reaction is easily determined from the heat of combustion of methane, we will use the extent of reaction method for the material and energy balance calculations.

$$CH_4 + 2O_2 \longrightarrow CO_2 + 2H_2O(v)$$

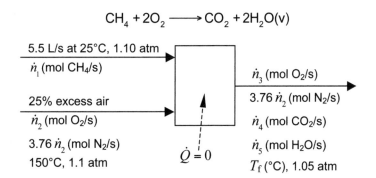

(9.66-1)

DEGREE-OF-FREEDOM ANALYSIS		
UNKNOWNS AND INFORMATION		JUSTIFICATION/CONCLUSION
+ 7 unknowns	_____	Use extent of reaction
− 4 extent of reaction equations	_____	
− 1 _____	_____	
− 1 _____	_____	
− 1 _____	_____	
0 DOF		Determine all unknowns

(9.66-2)

Q: Why don't we include an extent of reaction equation or a material balance for nitrogen?

A: _____

Do no arithmetic in the equations that follow—as before, we'll let E-Z Solve do it all.

Fuel feed rate

$$\dot{n}_1 = \frac{5.50\ \text{L}}{\text{s}} \left| \frac{\underline{\quad}\ \text{K}}{\underline{\quad}\ \text{K}} \right| \frac{\underline{\quad}\ \text{atm}}{\underline{\quad}\ \text{atm}} \left| \frac{\text{mol}}{22.4\ \text{L(STP)}} \right.$$

(9.66-3)

Oxygen feed rate

$$\dot{n}_2 = \frac{\dot{n}_1\ (\text{mol CH}_4\ \text{fed})}{\text{s}} \left| \frac{\underline{\quad}\ \text{mol O}_2\ \text{required}}{\text{mol CH}_4} \right| \frac{\underline{\quad}\ \text{mol O}_2\ \text{fed}}{\text{mol O}_2\ \text{required}}$$

(9.66-4)

Extent of reaction equations [Eq. (4.6-3), p. 119 in the text: $(\dot{n}_i)^{\text{out}} = (\dot{n}_{i0})^{\text{in}} + v_i\dot{\xi}$]

CH$_4$: $(v_i = -1)$: $\underline{\quad} = \underline{\quad} - \dot{\xi}$ (9.66-5)

O$_2$: _____   (9.66-6)

CO$_2$: _____   (9.66-7)

H$_2$O: _____   (9.66-8)

Dew point

The dew point can now be determined from Raoult's law, using the Antoine equation for the vapor pressure of water. (We could also determine the value of the vapor pressure and look up the temperature in Table B.3, but we want to do it all with E-Z Solve.)

$$y_{\text{H}_2\text{O}}P = p^*_{\text{H}_2\text{O}}(T_{dp}) \Rightarrow \frac{\dot{n}_5}{(\dot{n}_3 + \underline{\quad\quad\quad})}(760\ \text{mm Hg}) = 10^{\left\{8.10765 - \frac{\overline{\quad\quad}}{(\quad\quad)}\right\}}$$

(9.66-9)

Enthalpy table and enthalpies

(9.66-10)

References: CH$_4$(g), O$_2$, N$_2$, CO$_2$, H$_2$O(v) @ 25°C, 1 atm

	$\dot{n}_{\text{in}}$ (mol/s)	$\hat{H}_{\text{in}}$ (kJ/mol)	$\dot{n}_{\text{out}}$ (mol/s)	$\hat{H}_{\text{out}}$ (kJ/mol)
CH$_4$	$\dot{n}_1$			
O$_2$	____	$\hat{H}_A$	____	$\hat{H}_C$
N$_2$	____	$\hat{H}_B$	____	$\hat{H}_D$
CO$_2$			____	$\hat{H}_E$
H$_2$O			____	$\hat{H}_F$

$$\hat{H}_A = \hat{H}(\text{O}_2, 150°\text{C}) \xrightarrow{\text{Table B.8}} \underline{\quad\quad}\ \text{kJ/mol}$$

(9.66-11)

$$\hat{H}_B = \hat{H}(\text{N}_2, 150°\text{C}) \xrightarrow{\text{Table}\underline{\quad}} \underline{\quad\quad}\ \text{kJ/mol}$$

(9.66-12)

$$\hat{H}_C\left(\frac{kJ}{mol}\right) = \int_{25}^{T_f}\left(C_p\right)_{O_2} dT$$

$$\xrightarrow{\text{Table B.2}} \int_{25}^{T_f}\left[0.02910 + 1.158\times10^{-5}T - 0.6076\times10^{-8}T^2 + 1.311\times10^{-12}T^3\right]dT$$

$$= 0.02910(T_f - 25) + \frac{1.158\times10^{-5}}{2}(\underline{\hspace{2cm}}) - \frac{0.6076\times10^{-8}}{\underline{\hspace{1cm}}}(\underline{\hspace{2cm}})$$

$$+ \frac{1.311\times10^{-12}}{\underline{\hspace{1cm}}}(\underline{\hspace{2cm}})$$

(9.66-13)

$$\hat{H}_D\left(\frac{kJ}{mol}\right) = \int_{25}^{T_s}\left(C_p\right)_{N_2} dT$$

$$\xrightarrow{\text{Table B.2}} \underline{\hspace{1.5cm}}(T_f - 25) + \underline{\underline{\hspace{1.5cm}}}(T_f^2 - 25^2) + \underline{\underline{\hspace{1.5cm}}}(T_f^3 - 25^3)$$

$$- \underline{\underline{\hspace{1.5cm}}}(T_f^4 - 25^4)$$

(9.66-14)

$$\hat{H}_E\left(\frac{kJ}{mol}\right) = \int_{\underline{\hspace{0.5cm}}}^{\underline{\hspace{0.5cm}}}\left(C_p\right)_{CO_2} dT$$

$$\xrightarrow{\text{Table B.2}} 0.03611(\underline{\hspace{2cm}}) + \frac{4.233\times10^{-5}}{2}(\underline{\hspace{2cm}}) - \frac{2.887\times10^{-8}}{3}(\underline{\hspace{2cm}})$$

$$+ \frac{7.464\times10^{-12}}{4}(\underline{\hspace{2cm}})$$

(9.66-15)

$$\hat{H}_F\left(\frac{kJ}{mol}\right) = \int_{25}^{T_f}\left(C_p\right)_{H_2O(v)} dT$$

$$\xrightarrow{\text{Table B.2}} 0.03346(T_f - 25) + \frac{0.688\times10^{-5}}{2}(T_f^2 - 25^2) + \frac{0.7604\times10^{-8}}{3}(T_f^3 - 25^3)$$

$$- \frac{3.593\times10^{-12}}{4}(T_f^4 - 25^4)$$

Heat of reaction

The reaction is $CH_4 + 2O_2 \rightarrow CO_2 + 2H_2O(v)$. The heat of combustion of methane is listed in Table B.1 as -890.36 kJ/mol, but Footnote i on p. 629 tells us that this value is based on liquid water as a product. The footnote also indicates that to determine the value with water vapor as the product (which is what we want), we must add $44.01n_w$ to the tabulated value, where n_w is the moles of water formed per mole of fuel (methane) burned. Therefore,

$$\Delta\hat{H}_r^\circ = -890.36 + (\underline{\hspace{1cm}})(44.01)$$

(9.66-16)

> **Q:** Why is it critical for water vapor to be the presumed product when the heat of reaction is determined?
>
> **A:** The process path we are following to determine $\Delta\dot{H}$ involves bringing the feed stream components from their entering conditions to the reference conditions ($\Delta\dot{H} = -\sum \dot{n}_{in}\hat{H}_{in}$), carrying out the reaction at 25°C ($\Delta\dot{H} = \dot{\xi}\Delta\hat{H}_r^\circ$), and bringing the products from the reference conditions to the final temperature ($\Delta\dot{H} = \sum \dot{n}_{out}\hat{H}_{out}$). For the sum of these three terms to equal the desired enthalpy change for the overall process, _____.

Energy balance

$$\dot{Q} = \Delta\dot{H} \Rightarrow 0 = \dot{\xi}\Delta\hat{H}_r^\circ + \sum_{out}\dot{n}_i\hat{H}_i - \sum_{in}\dot{n}_i\hat{H}_i$$

(9.66-18)

$$\Rightarrow 0 = \dot{\xi}\Delta\hat{H}_r^\circ + \underline{\hspace{6cm}}$$

Degrees of superheat of product gas

DS = _____

(9.66-19)

E-Z Solve program

```
// Problem 9–66
// Mol balances (ni in mol/s)
n1 = 5.5*(273/298)*1.1/22.4        //Methane feed rate (Eq. 9.66–3)
n2 = n1*2*1.25                     //Oxygen feed rate (Eq. 9.66–4)

_____                      //Methane balance (Eq. 9.66–5)

_____                      //Oxygen balance (Eq. 9.66–6)

_____                      //CO2 balance (Eq. 9.66–7)

_____                      //H2O balance (Eq. 9.66–8)

760*n5/(n3+_____) = 10^(8.10765–1750.286/(Tdp+235))   //Raoult's law (Eq. 9.66–9)


// Enthalpies (H in kJ/mol)
HA = 3.78 ; HB = 3.66
HC = 0.02910*(Tf–25) + 1.158e–5*(Tf^2–25^2)/2 – 0.6076e–8*(Tf^3–25^3)/3 + 1.311e–12*(Tf^4–25^4)/4

HD = _____

HE = _____

HF = 0.03346*(Tf–25) + 0.6880e–5*(Tf^2–25^2)/2 + 0.7604e–8*(Tf^3–25^3)/3 – 3.593e–12*(Tf^4–25^4)/4

// Energy balance

DHr = _____

0 = Xi*DHr + n3*HC + 3.76*n2*HD + n4*HE + n5*HF – n2*HA – 3.76*n2*HB

DS = Tf–Tdp
```

E-Z Solve solution (Record your values.)

Feed	$\dot{n}_1 = $ _____ mol CH_4/s $\dot{n}_2 = $ _____ mol O_2/s
Effluent	$\dot{n}_3 = $ _____ mol O_2/s $\dot{n}_4 = $ _____ mol CO_2/s
	$\dot{n}_5 = 0.495$ mol $H_2O(v)$/s
Extent of reaction	$\dot{\xi} = $ _____ mol/s
Effluent temperature	$T_f = 1832°C$
Dew point	$T_{dp} = $ _____ °C
Degrees of superheat	DS = _____ °C

Notes and Calculations

Selected Answers

Chapter 2 – Introduction to Engineering Calculations

2.6

$$14,500 + 0.044643X = 21,500 + 0.027968X \implies X = \underline{420,000 \text{ miles}} \tag{2.6-2}$$

2.9

$$W(\text{lb}_f) = 40,000 \text{ lb}_f \tag{2.9-1}$$

2.25

$$r' = 84.7D' - 107.5 \, (D')^2 \tag{2.25-4}$$

2.31

$$Y = \underline{\sin y}, \ X = (\cos x)^2, \ s = a, \ I = b \tag{2.31-1}$$
$$Y = pC, \ I = k_2 \tag{2.31-2}$$
$$X = 1/p^2 \tag{2.31-4}$$
$$X = 1/V, \ s = a, \ I = b \tag{2.31-5}$$
$$I = 0 \tag{2.31-6}$$
$$Y = \underline{\ln C}, \ I = \ln C_0 \tag{2.31-8}$$
$$Y = \underline{C} \tag{2.31-9}$$
$$X = 1/T \tag{2.31-10}$$
$$Y = p, \ s = -\Delta H_v / R \tag{2.31-11}$$
$$Y = \ln(Q-4) \tag{2.31-12}$$
$$X = \ln(r^2-4) \tag{2.31-14}$$
$$(\text{semilog}) \ Y = y^3-3, \ s = B \tag{2.31-17}$$

2.32

$$y = aR + b \implies \left. \begin{array}{l} a = \dfrac{0.169 - 0.011}{80 - 5} = 2.11 \times 10^{-3} \\[2mm] b = 0.011 - \left(2.11 \times 10^{-3}\right)(5) = 4.50 \times 10^{-4} \end{array} \right\} \implies y = 2.11 \times 10^{-3} R + 4.50 \times 10^{-4} \tag{2.32-1}$$

$$R = 43 \implies \dot{m} = 110 \text{ kg H}_2\text{O/h} \tag{2.32-2}$$

2.38

E-Z Solve program (2.38-1)

```
// Problem 2.38(a)
ln(2.58) = lna + b*ln(1.02) + c*ln(11.2)        // "lna" is a variable name & represents ln(a)
ln(3.72) = lna + b*ln(1.75) + c*ln(11.2)
ln(3.5) = lna + b*ln(1.02) + c*ln(9.1)
a = exp(lna)
```

2.38 (con't)

E-Z Solve program (2.38-3)

```
// Problem 2.38(c)
//Initialize values of the dependent variables
Z1 = 2.27; Z2 = 2.58; Z3 = 3.72; Z4 = 5.21; Z5 = 3.50; Z6 = 4.19; Z7 = 5.89
//Force E-Z Solve to regress for values of lna, b, and c
ln(Z1) = lna + b*ln(0.65) + c*ln(11.2)
ln(Z2) = lna + b*ln(1.02) + c*ln(11.2)
ln(Z3) = lna + b*ln(1.75) + c*ln(11.2)
ln(Z4) = lna + b*ln(3.43) + c*ln(11.2)
ln(Z5) = lna + b*ln(1.02) + c*ln(9.1)
ln(Z6) = lna + b*ln(1.02) + c*ln(7.6)
ln(Z7) = lna + b*ln(1.02) + c*ln(5.4)
//
a = exp(lna)          //find a from the regressed variable lna
//
//Now calculate the auxilliary variables for comparison to the input Z values
Zcalc1 = a*0.65^b*11.2^c
Zcalc2= a*1.02^b*11.2^c
Zcalc3 = a*1.75^b*11.2^c
Zcalc4 = a*3.43^b*11.2^c
Zcalc5 = a*1.02^b*9.1^c
Zcalc6 = a*1.02^b*7.6^c
Zcalc7 = a*1.02^b*5.4^c
```

Chapter 3 – Processes and Process Variables

3.3

$$\text{Volumetric feed ratio} = \frac{V_g(\text{cm}^3 \text{ gasoline})}{1.00 \text{ cm}^3 \text{ kerosene}} = 0.5 \text{ cm}^3 \text{ gasoline/cm}^3 \text{ kerosene.} \qquad \text{(3.3-6)}$$

3.4

In U.S.: $\underline{\$36.80}$ (3.4-2)

3.10

$$\rho_{\text{bulk}} = \underline{\underline{2.05 \text{ kg/L}}} \qquad \text{(3.10-1)}$$

$$W_{bag} = \underline{\underline{1.00 \times 10^3 \text{ N}}} \qquad \text{(3.10-2)}$$

3.17

$$\dot{m}_{N_2} = \underline{\underline{2470 \text{ kg N}_2/\text{h}}} \qquad \text{(3.17-2)}$$

3.29

Condensate flow rate

$$\dot{n}_3 = 11.47 \, \frac{\text{mol C}_6\text{H}_{14}(l)}{\text{min}} \qquad \text{(3.29-1)}$$

Hexane Balance

$$0.18 \frac{\text{mol C}_6\text{H}_{14}}{\text{mol feed}} \times \dot{n}_1 \left(\frac{\text{mol feed}}{\text{min}} \right) = 0.05 \frac{\text{mol C}_6\text{H}_{14}}{\text{mol exit gas}} \times \dot{n}_2 \left(\frac{\text{mol exit gas}}{\text{min}} \right) + 11.47 \frac{\text{mol C}_6\text{H}_{14}(l)}{\text{min}} \qquad \text{(3.29-2)}$$

E-Z Solve solution

$$\dot{n}_2 = 72.3 \text{ mol/min} \qquad \text{(3.29-4)}$$

Hexane recovery = $\underline{\underline{76\%}}$ (3.29-5)

3.42

$$R = \frac{\rho_T(500 - h)}{\rho_M - \rho_T} \qquad \text{(3.42-2)}$$

Manometer fluid is mercury $\Rightarrow R_{\text{Hg}} = 23.9 \text{ cm}$ (3.42-4)

3.47

$$\Delta P \ \left(\text{mm Hg}\right) = \underline{0.0227 \ \text{h}\left(\text{mm}\right)} \qquad\qquad \textbf{(3.47-1)}$$

$$\ln \dot{v} = \underline{n \ln(\Delta P)} + \underline{\ln K} \qquad\qquad \textbf{(3.47-3)}$$

Plot $\mathbf{Y} = \ln(\dot{v})$ vs. $\mathbf{X} = \ln(\Delta P)$ on linear coordinates
Slope $= n$, Intercept $= \ln(K)$ $\qquad\qquad$ **(3.47-4)**

$$n = 0.4975 \ (\sim 0.5), \quad K = \exp(5.2065) = 182.45 \ \text{mL/s·cm Hg}^{0.4975} \qquad \textbf{(3.47-7)}$$

$$\dot{m} = 104.4 \ \frac{\text{g}}{\text{s}} \quad \dot{n} = 2.4 \frac{\text{mol}}{\text{s}} \qquad\qquad \textbf{(3.47-10)}$$

3.51

$$K = 3.1703 \, ^\circ\text{F} \qquad\qquad \textbf{(3.51-3)}$$
$$R = 49.3 \qquad\qquad \textbf{(3.51-5)}$$

Chapter 4 – Fundamentals of Material Balances

4.11

(4.11-1)

> Much higher rates of dilution air than the minimum would be used to make *certain* that the mixture remains outside the explosive region.

4.16

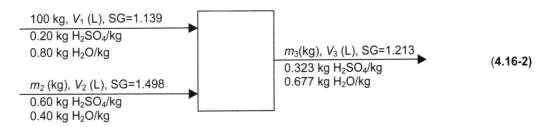

100 kg, V_1 (L), SG=1.139
0.20 kg H_2SO_4/kg
0.80 kg H_2O/kg

m_2 (kg), V_2 (L), SG=1.498
0.60 kg H_2SO_4/kg
0.40 kg H_2O/kg

m_3(kg), V_3 (L), SG=1.213
0.323 kg H_2SO_4/kg
0.677 kg H_2O/kg

(4.16-2)

$m_3 = 144$ kg
$V_3 = 118.7$ L product soln.

(4.16-5)

4.21

$$\dot{m}_B = 400 \ \text{lb}_m /\text{h}$$

(4.21-5)

$$x\% = 6.841e^{0.2682 R_x}$$

(4.21-9)

$$0.01(6.84e^{0.2682 R_x})(7.778R_A - 44.444) + 0.9(15.0R_B - 100) =$$
$$0.75(7.778R_A - 44.444 + 15.0R_B - 100)$$

(4.21-15)

4.32

E-Z Solve code

```
// Problem 4.32
// Simultaneous solution of process mass balances
100*0.12 = 0.42*m5              //overall sugar balance
100 = m3 + m5                   //overall mass balance
m4 + m2 = m5                    //mixing point total mass balance
0.58*m4 + 0.12*m2 = 0.42*m5     //mixing point sugar balance
100 = m1 + m2                   //bypass split total mass balance
```

$m_1 = 90.1$ kg fresh juice fed to evaporator

(4.32-5)

4.36

$$m_2 = 400 \text{ kg} \quad x_2 = 0.218 \text{ kg S/kg} \quad y_2 = 0.033 \text{ kg oil/kg}$$

(4.36-4)

$$\text{Recycle ratio} = \frac{m_5}{m_1} = 9.79 \text{ kg } C_6H_{14} \text{ recycled/kg } C_6H_{14} \text{ fed}$$

(4.36-5)

4.50

(4.50-10)

	n1	n2	n3	n4	n5
16:20	INS		Equation set successfully solved.		
1	714.286	114.286	607.143	114.286	7.14286

Fractional yield: $\dfrac{n_{C_2H_5Cl} \text{ actual}}{n_{C_2H_5Cl} \text{ maximum}} = 0.88$

(4.50-12)

4.59

(4.59-11)

(i) molar flow rates of ethylene and oxygen in the fresh feed

$\dot{n}_E = 15$ mol C_2H_4/h $\dot{n}_O = 11.25$ mol O_2/h

(ii) production rate of ethylene oxide

$\dot{n}_3 = 13.5$ mol C_2H_4O/h

Oxygen feed rate required to produce 1 ton/h C_2H_4O

(4.59-4)

$\left(\dfrac{2000}{44.05}\right)$ lb-mol C_2H_4O/h $\times \dfrac{\dot{n}_O \text{ mol } C_2H_4/h}{\dot{n}_3 \text{ mol } C_2H_4O/h} = \left(\dfrac{2000}{44.05}\right)\left(\dfrac{11.25}{13.5}\right)$

$= \underline{\underline{37.8 \text{ lb-mol } O_2/h \text{ required}}}$

4.70

O balance #1 45.36 mol O $\neq$ 39.72 mol O

(4.70-2)

O balance #2 44.4 mol O = 44.4 mol O

(4.70-4)

18.6% excess air

(4.70-6)

Chapter 5 – Single Phase Systems

5.10

$$\dot{n}_o = 29.883 \text{ mol/s} \tag{5.10-2}$$

5.19

$$\frac{RT}{P} = 24.4 \frac{L}{mol} \tag{5.19-2}$$

5.23

Weight of He in balloon = 20,250 N — (5.23-1)
Mass of air displaced = 1988 kg — (5.23-3)
Cable restraining force $= F_{buoyant} - w_{balloon} - w_{He} = 27,200$ N — (5.23-4)
Acceleration $(m/s^2) = 12.3$ m/s^2 — (5.23-5)

5.38

$$P_{final} = 27.3 \text{ atm} \tag{5.38-2}$$

5.46

E-Z Solve program (5.46-8)

```
// Problem 5.46
n0 = 50.4*0.630/72.05
n1 = n0*1.15*8/0.210
Vair= n1*1000*0.08206*336*101.325/(208.6+101.325)
n2 = 3.175/72.15
Xp = (n0-n2)/n0
n0*5 = n2*5 + n4
n0*12 = n2*12 + n5*2
0.210*n1*2 = n3*2 + n4*2 + n5
Vgas = (n3 + 0.790*n1 + n4)*1000*0.08206*275
Vcond = (3.175/0.630) + n5*18.02
```

5.62

E-Z Solve program (5.62-5)

```
// Problem 5.62(b)
P = (R*T)/(V–b)-alpha*a/(V*(V+b))     // SRK equation, psia
a = 0.42748*(R*Tc)^2/Pc               // ft6-psi/lbmol2
b = 0.08664*R*Tc/Pc                   // ft3/lbmol
m = 0.48508 + (1.55171*w) – 0.1561*w^2
alpha = (1+m*(1–sqrt(Tr)))^2
T = 509.7; P = 2417                   // temp,R & pressure, psia
Tc = 277.9; Pc = 730.4                // crit. temp., R, & pressure, psia (Tab. B.1)
w = 0.021                             // acentric factor (Tab. 5.3-1)
R = 10.73                             // ft3-psia/lb-mol-deg.R
Tr = T/Tc                             // reduced temp
//Initial guess for V = 2.3 ft^3/lb-mole
```

$$m_{O_2} = 37.5 \text{ lb}_m \tag{5.62-6}$$

5.68

74 mol O_2 obtainable   **(5.68-7)**

5.80

$$\dot{V}_{max} = 0.188 \frac{m^3}{min}$$ **(5.80-2)**

$$T_c' = 274.8 \text{ K}$$ **(5.80-3)**

$$T_r' = 1.32$$ **(5.80-4)**

Chapter 6 – Multiphase Systems

6.11

$$T_{dp} = \underline{38.1°C} \tag{6.11-1}$$

$$n_w = 1.34 \text{ cm}^3 \tag{6.11-3}$$

6.18

(6.18-2)

> As with all condensation processes, the exiting streams involve liquid in equilibrium with vapor. The water vapor at the outlet of the compressor is therefore saturated at the outlet temperature and pressure, and so Raoult's law (Eq. 6.3-1, p. 249 in the text) provides the fourth equation.

$$h_r = 1.75\% \tag{6.18-8}$$

$$0.310 \text{ m}^3 \text{ outlet air}/\text{m}^3 \text{ feed air} \tag{6.18-11}$$

6.24

(6.24-4)

> **Q:** Is it possible for the average gas phase composition in the tank (that is, the total nonane in the tank divided by the total moles of gas) to be within the explosion limits at any time?
>
> **A:** Yes. Right after the tank is drained, the gas in it initially contains 0% nonane (by our initial assumption). If all of the liquid evaporates and a negligible amount escapes, the gas contains 10% nonane. It follows that during the evaporation process the average composition passes through the explosive region.
>
> **Q:** Even if the maximum mole percent of nonane were below the lower explosive limit, why might explosion still be a danger? (Again, think about the assumptions.)
>
> **A:** The nonane will not spread uniformly—there will always be a region where the mixture is explosive at some time during the evaporation. (State where it would be.)

$$p^* = \exp\left(19.23 - \frac{5269}{T(\text{K})}\right) \tag{6.24-5}$$

$$T = 29°C \tag{6.24-6}$$

6.32

Cooler inlet $\xrightarrow{\text{Fig. 5.4-1}} z_{in} = 1.02$ (6.32-3)

Cooler outlet $\xrightarrow{\text{Fig. 5.4-3}} z_{out} = 0.98$

$$\dot{n}_F = 3.95 \times 10^4 \text{ mol/min} \tag{6.32-4}$$

$$\dot{V}_F = 25.7 \text{ m}^3/\text{min} \tag{6.32-5}$$

$$\dot{n}_G = 2.97 \times 10^4 \text{ mol/min}$$

$$\Rightarrow \dot{n}_{CH_3OH} = 14.8 \text{ mol } CH_3OH/\text{min} \tag{6.32-10}$$

6.39

$$y_w = \frac{77.60 \text{ mm Hg}}{780 \text{ mm Hg}} \tag{6.39-2}$$

$$y_w = \frac{n_w}{100 + n_w} \Rightarrow n_w = \frac{100 * y_w}{1 - y_w} \tag{6.39-3}$$

$$3n_p + 4n_b = (100)\left[(0.000527)(3) + (0.000527)(4) + 0.0148 + 0.0712\right] \tag{6.39-4}$$

6.54

$$p_P^* \approx 140 \text{ psi}, \ p_{nB}^* \approx 35 \text{ psi}, \ p_{iB}^* \approx 50 \text{ psi} \quad \text{(Answers may vary considerably)} \tag{6.54-4}$$

77% propane in head space. (Show the calculation in the space provided.) $\qquad$ (6.54-7)

6.60

$$\left(85.0 \ \frac{\text{mol}}{\text{h}}\right)\left(0.980 \frac{\text{mol A}}{\text{mol}}\right) = 0.950(0.450\dot{n}_0) \Rightarrow \dot{n}_0 = 195 \text{ kmol/h} \tag{6.60-2}$$

$$T_{dp} = 37.3°C \tag{6.60-8}$$

$$\dot{V}_{\text{distillate}} = 9.9 \times 10^3 \text{ L/h} \tag{6.60-12}$$

(6.60-15)

```
//E-Z Solve Program for Problem 6-60
//Part (b)
xa = 0.980*760/10^(6.84471–1060.793/(Tdp+231.541))
xb = 0.020*760/10^(6.88555–1175.817/(Tdp+224.867))
xa + xb = 1
//Part (c)
ya = 0.0405*10^(6.84471–1060.793/(Trb+231.541))/760
yb = 0.9595*10^(6.88555–1175.817/(Trb+224.867))/760
ya + yb = 1
```

(6.60-17)

A: The two equations are equivalent. [You can get Eq. (6.60-14) by dividing Eq. (6.4-4) by the total pressure P (= 760 mm Hg).] The conclusion is that the reboiler temperature is the same as the bubble point temperature of the bottoms product stream. When heated at 1 atm, a liquid mixture of that composition below 66.6°C would form a first bubble at 66.6°C. At that temperature, the vapor in the bubble would contain 10.3 mole% pentane and 89.7 mole% hexane because this is the only vapor composition that can coexist in equilibrium with the specified liquid composition at 1 atm. Similarly, 66.6°C is the dew point temperature at 1 atm of a vapor with that composition.

Chapter 7 – Energy and Energy Balances

7.21

$T_{ref} = 25°C$ (Show your work) (7.21-1)

Average rate of heat removal = 20.8 kW (7. 21-4)

7.28

$\dot{Q} = 5.47 \times 10^3$ kW (7. 28-4)

$\dot{V}_{steam} = 1.27$ m^3/s (7.28-6)

7.41

Raoult's law at outlet $\overbrace{x_1}P_1 = p_W^*(38°C)$ [7.41-2] (7. 41-2)

1 atm   Table B.3

$x_1 = 0.0204$ kmol H$_2$O/kmol (7. 41-4)

$\dot{m}_2 = 17.7$ kg/min H$_2$O condenses (7. 41-7)

Table B.5

$\hat{H}_{H_2O}(v, 38°C) = \dfrac{2570.8 \text{ kJ}}{\text{kg}} \left| \dfrac{1 \text{ kg}}{10^3 \text{ g}} \right| \dfrac{18.02 \text{ g}}{\text{mol}} = 46.33$ kJ/mol (7. 41-8)

$\dot{Q} = 270$ tons of cooling (7. 41-9)

7.45

$T_{feed} = 212.4°C$ (State where this comes from.) (7.45-1)

Inlet vapor: $\hat{H} = 2797.2$ kJ/kg (from Table B.7) (7.45-3)

$\hat{H}_{ad} = 2741$ kJ/kg (7. 45-4)

(7. 45-6)

As the steam (which is transparent) moves away from the trap, it cools. When the steam reaches its saturation temperature at 1 atm, it begins to condense, so that $T = 100°C$.

Chapter 8 – Balances on Nonreactive Processes

8.9

$$\dot{Q} = 17,650 \text{ kW}$$ (8.9-3)

$$Q = 13,500 \text{ kJ}$$ (8.9-7)

8.18

(8.18-2)

> $\hat{H}_2$ is the specific enthalpy of $C_6H_{14}O(l,\underline{45}°C)$ relative to $C_6H_{14}O(l,\underline{20}°C)$
>
> $\therefore$ $\hat{H}_2$ is $\Delta\hat{H}$ for the process $C_6H_{14}O(l,\underline{20}°C) \rightarrow C_6H_{14}O(l,\underline{45}°C)$
>
> $$\hat{H}_2 = \int_{20°C}^{45°C} (C_p)_{C_6H_{14}O(l)} dT$$
>
> The specific enthalpies $\hat{H}_1$ and $\hat{H}_4$ equal zero, because they refer to species at their reference states. (They can be entered directly into the table.)

$$\hat{H}_3 = -3.226 \text{ kJ/mol} = -54.6 \text{ kJ/kg}$$ (8.18-3)

$$(C_p)_{C_6H_{14}O(l)} \approx 0.349 \frac{\text{kJ}}{\text{mol} \cdot °C} \Rightarrow \hat{H}_2 = 85.5 \frac{\text{kJ}}{\text{kg}}$$ (8.18-4)

$$\dot{Q} = \underline{-572 \text{ kW}} \text{ (572 kW of cooling)}$$ (8.18-6)

8.22

(8.22-1)

> - Do the degree-of-freedom analysis on the furnace to verify that the stack gas component amounts (n_2, n_3, n_4, n_5) can be determined. If they can, determine them.
> - Write an energy balance on the gas side of the boiler to determine Q. [Solves Part (a).]
> - Write an energy balance on the steam side of the boiler to determine m_w. [Solves Part (b).]
> - Scale up the process to the specified steam production rate and determine the required fuel feed rate. [Solves Part (c).]

$$Q = -21.2 \times 10^3 \text{ kJ}$$ (8.22-10)

$$m_w = 8.11 \text{ kg steam produced}$$ (8.22-11)

$$\dot{V}_{stack} = 3.40 \text{ m}^3 / \text{s}$$ (8.22-14)

8.30

$$\hat{H}_a = (\underline{104.8} - \underline{3488}) \frac{\text{kJ}}{\text{kg}} \times \left(\frac{18.02 \text{ kg}}{\text{kmol}} \right) \times \left(\frac{1 \text{ kmol}}{10^3 \text{ mol}} \right)$$ (8.30-8)

$$\dot{n}_3(1 - x_3)\hat{H}_b + \dot{n}_3 x_3 \hat{H}_c - \dot{n}_2 \hat{H}_a = 0$$ (8.30-11)

$$T_{ad} = 138°C$$ (8.30-13)

$$\dot{Q} = -292 \text{ kW}$$ (8.30-14)

8.46

Raoult's law for inlet air: $(x_1)(760 \text{ torr}) = 83.71$ torr $\qquad$ (8.46-3)

0.0606 kg H_2O condensed/m^3 air fed $\qquad$ (8.46-8)

12.6 kW of cooling $\qquad$ (8.46-9)

8.52

(8.52-3)

> **Q:** Not counting the reference states and column heading row, how many rows will the table need, and why?
> **A:** <u>Four.</u> There are two different process states at the inlet (liquid benzene at 25°C and liquid toluene at 25°C) and four at the outlet (liquid and vapor benzene, liquid and vapor toluene, all at 95°C). We will need rows for B(l), T(l), B(v), and T(v).

(8.52-4)

> **Q:** Since we are not going to use tabulated enthalpy data, the choice of reference states is arbitrary. What are convenient choices for benzene? What makes them convenient?
> **A:** The three states (phases and temperatures) of benzene in the inlet and outlet streams at 1 atm: <u>B(l, 25°C, 1 atm), B(l, 95°C, 1 atm), B(v, 95°C, 1 atm).</u> (<u>We have the same choices for toluene.</u>) They are convenient because they each allow us to set one value of $\hat{H}_i$ in the table equal to 0 rather than having to calculate all three. We choose 1 atm because that is the pressure for which the heat capacities in Table B.2 and the heats of vaporization in Table B.1 are given, and we ignore the effect of pressure on specific enthalpy when we calculate $\hat{H}_i$.

$\hat{H}_1 = \int_{95°C}^{25°C} \left(C_p\right)_{B(l)} dT = -9.838$ kJ/mol $\qquad$ (8.52-6)

$T(l, 95°C) \rightarrow T(l, 25°C) \qquad \hat{H}_2 = -11.78$ kJ/mol $\qquad$ (8.52-7)

$\hat{H}_3 = 30.08$ kJ/mol $\qquad$ (8.52-9)

$\dot{Q} = 2.42 \times 10^4$ kW $\qquad$ (8.52-11)

$p_T = 274$ torr $\qquad$ (8.52-13)

$y_B = 0.646$ mol B(v)/mol $\qquad$ (8.52-14)

8.72

(8.72-1)

> **Q:** What other property of the air that can be located on the psychrometric chart is given in the process statement?
> **A:** We are told that if the gas is cooled at (essentially) constant pressure, condensation begins at 20°C. By definition, $T_{\text{dew point}} = 20°C$.

$T_{\text{wet bulb}} \approx 25.5°C, \quad V_h \approx 0.908$ m^3/kg DA. $\qquad$ (8.72-2)

8.72 (con't)

$$m_{H_2O} = 3.26 \times 10^{-5} \text{ kg } H_2O(v) \tag{8.72-4}$$

$$\hat{H}_{initial} = 78.3 \text{ kJ/kg DA (Answers may vary slightly)} \tag{8.72-5}$$

$$\Delta H = -45 \text{ J} \tag{8.72-6}$$

$$Q = -32 \text{ J} \tag{8.72-7}$$

8.86

$$\tag{8.86-1}$$

- Calculate m_S (g solution) from the volume (1 liter) and specific gravity
- Calculate n_H (mol HCl) from the volume and molarity of the solution
- Calculate m_H (g HCl) from n_H and the molecular weight of HCl
- Calculate m_W (g H_2O) as $m_S - m_H$ and n_W from m_W and the molecular weight of H_2O
- Calculate $V_H = n_H RT/P$

$$n_W = 46.0 \text{ mol } H_2O, \ V_H = 185 \text{ L HCl(g)} \tag{8.86-2}$$

$$\hat{H}_1 = -0.15 \text{ kJ/mol} \tag{8.86-3}$$

$$\hat{H}_2 = -59.1 \text{ kJ/mol HCl} \tag{8.86-4}$$

$$Q = -470.4 \text{ kJ} \tag{8.86-5}$$

$$T_{ad} = 192.5°C \tag{8.86-8}$$

9.2

$$\dot{Q} = \underline{-1.53 \times 10^5} \text{ kW} \tag{9.2-3}$$

$$\tag{9.2-4}$$

We assumed that the reactor pressure is low enough to have a negligible effect on enthalpy.

$$\tag{9.2-6}$$

Yes. Pure n-nonane can only exist as vapor at 1 atm above 150.6°C, but it can exist as a vapor at lower temperatures in a mixture of gases. (In the same way, pure water can only exist as vapor at 1 atm above 100°C, but it can exist as a vapor at lower temperatures in a mixture of gases. In fact, you're breathing some right now.)

9.10

$$\text{Table B.2} \Rightarrow (C_p)_{H_2O(l)} = \underline{75.4 \times 10^{-3}} \text{ kJ/mol} \cdot °C \tag{9.10-2}$$

$$Q = 89.4 \text{ kJ} \tag{9.10-3}$$

$$\Delta \hat{H}_c^\circ = -5068 \text{ kJ/mol} \tag{9.10-5}$$

9.16

$$\dot{n}_1 = \dot{n}_0 \tag{9.16-3}$$

$$\dot{V}_0 \text{ (SCMS)} = \frac{\dot{n}_0(22.4)}{10^3} \tag{9.16-8}$$

$$\Delta \hat{H}_r^\circ = -98.28 \text{ kJ/mol} \tag{9.16-10}$$

$$\tag{9.16-15}$$

Q: Why did we put a minus sign next to $\dot{Q}$ on the flowchart?
A: Because whatever the sign of $\dot{Q}$ in the energy balance of Part (b), it must be the opposite in the energy balance on the cooling jacket. [In fact, it must be negative in Part (b) since heat is being transferred from the reactor to the coolant, raising the coolant's temperature.]

$$\dot{n}_0 = 32.0 \text{ mol/s (SO}_2 \text{ fed)} \qquad \dot{V}_1 = \underline{3.41 \text{ SCMS}} \text{ (air fed)} \tag{9.16-18}$$

9.21

$$n_1 = 0.5102 \text{ mol C}_2\text{H}_4 \tag{9.21-3}$$

$$n_3 = 0.02417 \text{ mol C}_2\text{H}_5\text{OH} \tag{9.21-4}$$

$$n_4 = 1.415 \times 10^{-3} \text{ mol } (C_2\text{H}_5)_2\text{O} \tag{9.21-5}$$

$$n_2 = 0.3414 \text{ mol H}_2\text{O} \tag{9.21-6}$$

$$2C(s,25°C) + 2H_2(g,25°C) \rightarrow C_2H_4(g,25°C) \rightarrow C_2H_4(g,310°C) \tag{9.21-9}$$

$$\hat{H}_1 = 68.69 \text{ kJ/mol} \tag{9.21-10}$$

$$\hat{H}_2 = 233.58 \text{ kJ/mol} \tag{9.21-11}$$

$$\hat{H}_3 = -211.15 \text{ kJ/mol} \tag{9.21-12}$$

9.33

$$\Delta \hat{H}_r^\circ = -90.68 \text{ kJ/mol} \tag{9.33-1}$$

$$\dot{n}_2 = 2331 - 2\dot{\xi} \tag{9.33-4}$$

$$\hat{H}_2 = 2.944 \text{ kJ/mol} \tag{9.33-7}$$

$$\dot{Q} = \Delta \dot{H} \overset{\text{Eq. (9.5-1a)}}{=} \dot{\xi}\Delta \hat{H}_r^\circ + \sum_{\text{out}} \dot{n}_i \hat{H}_i - \sum_{\text{in}} \dot{n}_i \hat{H}_i$$

$$\Rightarrow \frac{-17.05 \text{ kJ}}{\text{s}} \left| \frac{3600 \text{ s}}{1 \text{ h}} \right. = \left(-90.68 \frac{\text{kJ}}{\text{mol}}\right)\left[\dot{\xi}\left(\frac{\text{mol}}{\text{s}}\right)\right] + \left[(1166 - \dot{\xi})\frac{\text{mol}}{\text{h}}\right]\left(2.987 \frac{\text{kJ}}{\text{mol}}\right) \tag{9.33-8}$$

$$+ (2331 - 2\dot{\xi})(2.944) + 5.009\dot{\xi}$$

$$\Rightarrow \dot{\xi} = 759 \text{ mol/h}$$

$$\dot{V}_{\text{out}} = 13.0 \text{ m}^3/\text{h} \tag{9.33-10}$$

9.45

Basis 1 mol H_2SO_4 fed $\xrightarrow{\text{2 mole\% solution}}$ 50 mol acid solution

$$\Rightarrow \underline{49} \text{ mol } H_2O$$

$$\tag{9.45-1}$$

$\xrightarrow{\text{acid is neutralized}}$ 2 mol NaOH fed $\xrightarrow{\text{5 mole\% solution}}$ 40 mol basic solution

$$\Rightarrow \underline{38} \text{ mol } H_2O$$

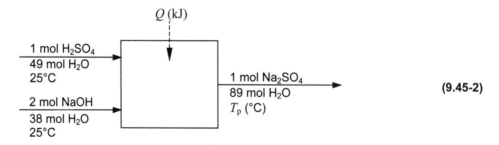

$$\tag{9.45-2}$$

The water in the product stream is the 87 moles that entered
the vessel plus the two moles formed in the reaction.

$$\tag{9.45-3}$$

References: $H_2(g)$, $S(s)$, $O_2(g)$, $Na(s)$ @ 25°C, 1 atm

	n_{in} (mol solute)	$\hat{H}_{\text{in}}$ (kJ/mol solute)	n_{out} (mol solute)	$\hat{H}_{\text{out}}$ (kJ/mol solute)
H_2SO_4 (aq)	1.0	$\hat{H}_1$		
NaOH (aq)	2.0	$\hat{H}_2$		
Na_2SO_4 (aq)			1.0	$\hat{H}_3$

9.45 (con't)

$$H_2(g) + S(s) + 2O_2(g) \longrightarrow H_2SO_4(l, 25°C)$$

$$\searrow \qquad H_2SO_4(aq, 25°C, r = 49) \qquad \text{(9.45-4)}$$

$$49H_2(g) + \frac{49}{2}O_2(g) \longrightarrow 49H_2O(l, 25°C)$$

$$\Delta\hat{H} = \hat{H}_1 = \left(\Delta\hat{H}_f^\circ\right)_{H_2SO_4(l)} + 49\left(\Delta\hat{H}_f^\circ\right)_{H_2O(l)} + \Delta\hat{H}_s^\circ\left(r = 49\right)$$

$$= [\underline{-811.32} + 49\left(\Delta\hat{H}_f^\circ\right)_{H_2O(l)} + \underline{(-73.3)}] \text{ kJ/mol } H_2SO_4 \qquad \text{(9.45-5)}$$

$$= \left[-884.6 + 49\left(\Delta\hat{H}_f^\circ\right)_{H_2O(l)}\right] \text{kJ/mol } H_2SO_4$$

$$\hat{H}_2 = \left[-469.4 + 19\left(\Delta\hat{H}_f^\circ\right)_{H_2O(l)}\right] \text{kJ/mol NaOH} \qquad \text{(9.45-7)}$$

$$\hat{H}_3 = \left[-1276 + 89\left(\Delta\hat{H}_f^\circ\right)_{H_2O(l)}\right] \text{kJ/mol } Na_2SO_4 \qquad \text{(9.45-10)}$$

$$\hat{H}_3 = \left[-1385.7 + 89\left(\Delta\hat{H}_f^\circ\right)_{H_2O(l)}\right] + 7.304(T_p - 25) \qquad \text{(9.45-12)}$$

$$T_p = 43.3°C \qquad \text{(9.45-13)}$$

9.55

$$\left(\dot{n}_o\right) = \frac{1000 \text{ mol } CH_4}{s} \left| \frac{2 \text{ mol } O_2 \text{ required}}{\text{mol } CH_4} \right| \frac{1.1 \text{ mol } O_2 \text{ fed}}{\text{mol } O_2 \text{ required}} \qquad \text{(9.55-3)}$$

$$\frac{1000 \text{ mol } CH_4}{s} \left| \frac{1 \text{ mol C}}{1 \text{ mol } CH_4} \right. = \dot{n}_1(1) + \left(\dot{n}_3\right)(1) + (10\dot{n}_3)(1) \qquad \text{(9.55-5)}$$

$$\hat{H}_B = \hat{H}(O_2, 100°C) \xrightarrow{\text{Table B.8}} 2.24 \text{ kJ/mol} \qquad \text{(9.55-10)}$$

$$\hat{H}_D = \left(\Delta\hat{H}_f^\circ\right)_{CH_4} + \int_{25}^{400} (C_p)_{CH_4} dT \xrightarrow[\text{Phys. Prop. Database}]{\text{Table B.1}} (\underline{-74.85} + \underline{17.23}) \text{ kJ/mol} \qquad \text{(9.55-12}$$

$$\left(\dot{Q}\right) = -5.85 \times 10^5 \text{ kW} \qquad \text{(9.55-19)}$$

9.55 (con't)

(9.55-20)

> (i) If T_{air} increases, less of the energy of combustion is required to bring the oxygen and nitrogen to the reaction temperature so that <u>more energy is transferred out of the reactor as heat</u>. In the calculations, the specific enthalpies of the oxygen and nitrogen in the feed would be greater, $\sum_{in} \dot{n}_i \hat{H}_i$ would accordingly be greater, and $\dot{Q}$ would therefore be more
>
> negative since that sum is subtracted in the energy balance.

9.66

Fuel feed rate $\qquad \dot{n}_1 = \dfrac{5.50 \text{ L}}{\text{s}} \left| \dfrac{273 \text{ K}}{298 \text{ K}} \right| \dfrac{1.1 \text{ atm}}{1.0 \text{ atm}} \left| \dfrac{\text{mol}}{22.4 \text{ L(STP)}} \right.$ (9.66-3)

$$0 = \dot{n}_1 - \dot{\xi} \qquad\qquad (9.66\text{-}5)$$

$$\hat{H}_A = \hat{H}(O_2, 150°C) \xrightarrow{\text{Table B.8}} 3.78 \text{ kJ/mol} \qquad\qquad (9.66\text{-}11)$$

$$
\begin{aligned}
\hat{H}_C \left(\frac{\text{kJ}}{\text{mol}}\right) &= \int_{25}^{T_f} \left(C_p\right)_{O_2} dT \\
&\xrightarrow{\text{Table B.2}} \int_{25}^{T_f} \left[0.02910 + 1.158 \times 10^{-5} T - 0.6076 \times 10^{-8} T^2 + 1.311 \times 10^{-12} T^3 \right] dT \\
&= 0.02910(T_f - 25) + \frac{1.158 \times 10^{-5}}{2}(T_f^2 - 25^2) - \frac{0.6076 \times 10^{-8}}{3}(T_f^3 - 25^3) \\
&\quad + \frac{1.311 \times 10^{-12}}{4}(T_f^4 - 25^4)
\end{aligned}
$$

(9.66-13)

$$\Delta\hat{H}_r^{\circ} = -890.36 + (2)(44.01) \qquad\qquad (9.55\text{-}16)$$

(9.66-17)

> **Q:** Why is it critical for water vapor to be the presumed product when the heat of reaction is determined?
>
> **A:** The process path we are following to determine $\Delta\dot{H}$ involves bringing the feed stream components from their entering conditions to the reference conditions ($\Delta\dot{H} = -\sum \dot{n}_{in} \hat{H}_{in}$), carrying out the reaction at 25°C ($\Delta\dot{H} = \dot{\xi}\Delta\hat{H}_r^{\circ}$), and bringing the products from the reference conditions to the final temperature ($\Delta\dot{H} = \sum \dot{n}_{out} \hat{H}_{out}$). For the sum of these three terms to equal the desired enthalpy change for the overall process, the reaction products from the second step must be the starting points for the third step—namely, gases and (for water) vapor at 25°C.

$$DS = T_f - T_{dp} \qquad\qquad (9.66\text{-}19)$$

9.66 (con't)

(9.66-20)

Feed	$\dot{n}_1 = 0.247$ mol CH_4/s, $\dot{n}_2 = 0.619$ mol O_2/s
Effluent	$\dot{n}_3 = 0.124$ mol O_2/s, $\dot{n}_4 = 0.247$ mol CO_2/s, $\dot{n}_5 = 0.495$ mol $H_2O(v)$/s
Extent of reaction	$\dot{\xi} = 0.247$ mol/s
Effluent temperature	$T_f = 1832°C$
Dew point	$T_{dp} = 55°C$
Degrees of superheat	$DS = 1832°C - 55°C = 1777°C$